Leckie

the education publisher for Scotland

National 5
BIOLOGY

For SQA 2019 and beyond

Revision + Practice
2 Books in 1

001/01102020

10 9 8 7 6 5 4 3 2 1

ISBN 9780008435349

Published by
Leckie
An imprint of HarperCollinsPublishers
Westerhill Road, Bishopbriggs, Glasgow, G64 2QT
T: 0844 576 8126
leckiescotland@harpercollins.co.uk   www.leckiescotland.co.uk

Publisher: Sarah Mitchell
Project Manager: Harley Griffiths and Lauren Murray

Special thanks to
QBS (layout and illustration)

Printed in Italy by Grafica Veneta S.p.A

A CIP Catalogue record for this book is available from the British Library.

Acknowledgements
We would like to thank the following for permission to reproduce their material:
Knorre (p.36, 78) F. JIMENEZ MECA (p.45), Jupiterimages (p.46, 73), Stockbyte (p.69, 70), Comstock Images (p.77), Anup Shah (p.69, 70), Ryan McVay (p.76), NIGEL CATTLIN/SCIENCE PHOTO LIBRARY (p.85), Tom Brakeeld (p.69), Chris Howes/Wild Places Photography/ Alamy (p.72), PHILIPPE PSAILA/SCIENCE PHOTO LIBRARY (p.72), Cosmin Manci (p.71), Creatas Images / Getty (p.95), Hemera Technologies (p.96), Nicku (p.97), MICHAEL W. TWEEDIE/ SCIENCE PHOTO LIBRARY (p.98), DR KEITH WHEELER/SCIENCE PHOTO LIBRARY (p.78), THOMAS AMES JR., VISUALS UNLIMITED / SCIENCE PHOTO LIBRARY (p.78), Cray Photo (p.109), WAYNE LAWLER/SCIENCE PHOTO LIBRARY (p.91), jirapong (p.93)

All other images © Shutterstock.com

Whilst every effort has been made to trace the copyright holders, in cases where this has been unsuccessful, or if any have inadvertently been overlooked, the Publishers would gladly receive any information enabling them to rectify any error or omission at the first opportunity.

ebook

To access the ebook version of this Revision Guide visit
www.collins.co.uk/ebooks
and follow the step-by-step instructions.

Contents

Contents

Area 3: Biology: Life on Earth

ANSWERS Check your answers to the practice exam papers online: www.leckieandleckie.co.uk

Introduction

Complete Revision and Practice

This Complete **two-in-one Revision and Practice** book is designed to support you as students of National 5 Biology. It can be used either in the classroom, for regular study and homework, or for exam revision. By combining **a revision guide and two full sets of practice exam papers**, this book includes everything you need to be fully familiar with the National 5 Biology exam. As well as including ALL the core course content with practice opportunities, there is comprehensive assignment and exam preparation advice, easy reference with a glossary and both revision question and pratice exam paper answers provided online at www.leckieandleckie.co.uk.

About the revision guide

This guide has been written primarily as a revision tool for the National 5 qualification in Biology. Topics are usually displayed in double-page spreads with many topics taking up several double-page spreads. Information is presented in 'bite-sized' chunks and illustrated with appropriate graphics to enhance your understanding and learning. Remember that Biology is a very visual science! Top tips will appear frequently. These are based on many years' experience teaching students not only *what* to learn but also *how* to learn in Biology. Topics have 'Quick Tests' with online answers. Each section has a more global set of questions called 'Revision Questions' which include problem-solving. There are multiple-choice style and short and longer response questions. Answers to these tests are also provided online. These are very powerful ways of giving feedback on how well you have mastered a section of the course. In addition, all eight of the 'scientific inquiry skills' specified by The Scottish Qualification Authority are tested. A comprehensive glossary is included and an excellent idea would be to make flash-cards of these terms using readily available and free software from the Internet.

With any revision programme, it is essential to make a good and early start and be self-disciplined enough to follow this through. Every student's study and revision strategies are different and it is important that you look carefully at how you learn. If your strategies are not producing the results you would expect, then revisit them and try out some new and different ideas. Talk these over with your teacher and friends. Engage with your learning and don't just 'read over your notes', one of the poorest forms of revision and consolidation. Critically, if you hit a problem, discuss this as soon as possible with your teacher to resolve it before it escalates into an issue which might impact seriously on your final award. Above all, be thorough!

About the practice exam papers

The two papers included in the practice exam papers are designed to provide practice in the National 5 Biology course assessment question paper (the examination), which is worth 80% of the final grade for this course.

Together, the two papers give overall and comprehensive coverage of the assessment of **knowledge and its application** as well as the **skills of scientific inquiry** needed to pass National 5 Biology.

The actual final examination paper assesses your knowledge-base across all of the areas of Biology. In addition, it will assess eight different skills that are, in general terms, inquiry, analytical thinking as well as the impact of applications on society and the environment. We recommend that candidates download a copy of the Course Specification from the SQA website at www.sqa.org.uk. Print pages 25–47, which summarise the knowledge and skills that will be tested.

The final question paper, lasting 2.5 hours, is worth 100 marks and contributes 80% of the total mark you can achieve. This paper has two sections:

Section 1 has multiple-choice questions and is worth 25 marks.

Section 2 has structured and extended-response questions and is worth 75 marks.

The examiners will ensure the marks available are distributed proportionally across the whole course which reinforces the need to have a global command of the syllabus.

Approximately 70 marks are for knowledge and understanding, and approximately 30 marks are for applying scientific inquiry skills.

An integral part of the assessment is the completion of an internal assignment which is worth 20 marks. This will be an individually produced piece of work at a relevant point in the course and under supervised conditions. The final report you will produce is marked externally by SQA.

Leckie
the education publisher
for Scotland

National 5
BIOLOGY
For SQA 2019 and beyond

Revision Guide

John Di Mambro

Cell structure

Basic cell structure

TOP TIP

When drawing diagrams in Biology, don't use arrowed-headed lines and make sure the lines end exactly on the structure you are labelling.

Under a light microscope, you can see that living things are made up of cells, which share similar structures: **cell membrane**, **cytoplasm** and **nucleus**. Green plant cells have additional features: **chloroplasts**, **vacuole** and **cell wall**.

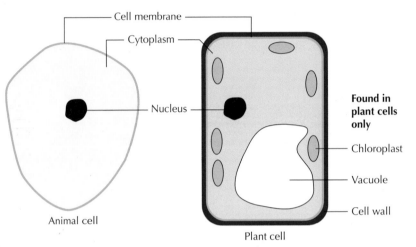

Cell membrane

Cytoplasm

Nucleus

Found in plant cells only

Chloroplast

Vacuole

Cell wall

Animal cell

Plant cell

Cell structure	Function
Cell membrane	Outer covering of cells that controls what can enter or leave
Cytoplasm	The watery substance found inside cells where all the chemical reactions of the cell take place
Nucleus	Controls all the activities of a cell and contains the genetic material
Chloroplast	Structure found in green plant cells that contains the pigment **chlorophyll** and where **photosynthesis** takes place
Vacuole	A membrane-bound sac found in plant cells containing a watery solution giving support
Cell wall	Outer covering of plant cells giving support and shape and is made of **cellulose**

Animal and plant cell ultrastructure

Under a different type of microscope which can reveal the cell in great detail, many more structures can be seen. Using a more powerful microscope the cell can be viewed in greater detail. Shown in the diagrams below are two additional structures visible under an electron microscope. Under the electron microscope we can see the cell **ultrastructure**.

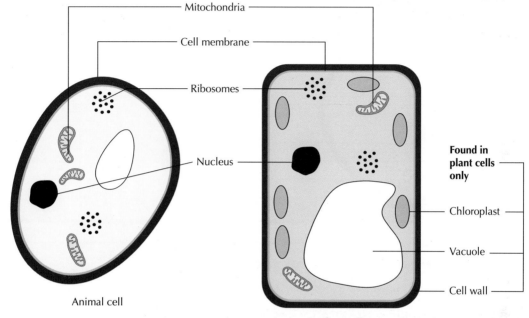

Animal cell

Plant cell

Cell structure	Function
Mitochondrion	Cylindrically-shaped structure found in varying numbers within the cytoplasm of cells that is the site of energy release in aerobic respiration
Ribosome	Small structure that is the site of protein synthesis in a cell

In multicellular organisms, cells can become specialised to perform particular functions. For example, the cells which line the cheek form a continuous layer of cells that are the same size and shape to form a protective lining. Cells in a leaf that contain chloroplasts are grouped together to allow for maximum exposure to sunlight, allowing photosynthesis to take place.

Fungal cell ultrastructure

Yeast is an example of a **fungus** that lives as a single-celled organism. Yeast cells are similar to plant cells, having a nucleus, cytoplasm, cell membrane, cell wall, vacuole, ribosomes and mitochondria, but they don't possess chloroplasts so can't photosynthesise. The cell wall has a different composition to that of plants and is not made of cellulose. Yeast can feed on dead animals and plants and so is very important in decomposition. Some yeasts are used in the manufacture of bread and alcohol.

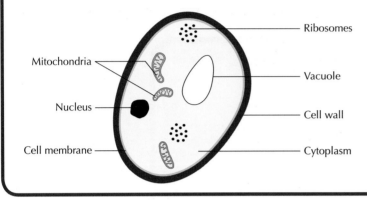

Bacterial cell ultrastructure

Bacteria are single-celled organisms that are usually very small. They share some features with other cells, having a cell wall, cytoplasm, cell membrane and ribosomes, but are also very different in having no nucleus or cell structures such as mitochondria, chloroplasts or vacuoles. In addition to their main genetic material, which is a large circular **chromosome**, they also possess smaller rings of genetic material called **plasmids**. These can reproduce independently of the main genetic material and can also pass between bacteria. Like fungi, bacterial cell walls are chemically different from plant cell walls. They are not made of cellulose.

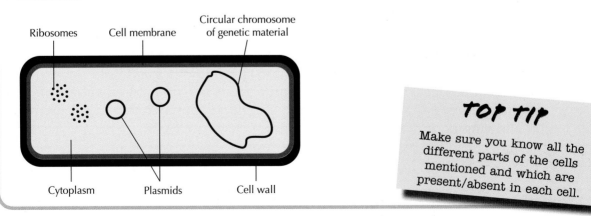

TOP TIP

Make sure you know all the different parts of the cells mentioned and which are present/absent in each cell.

Measuring cell size

Since cells are so small, a special unit of measurement is used. This is called the **micron** (or micrometre) and is often indicated by the symbol μm:

$$1 \text{ μm} = 0\cdot001 \text{ mm}$$
$$1 \text{ mm} = 1000 \text{ μm}$$

If a single-celled animal was 1·5 mm long and 0·5 mm wide, its size would be expressed as 1500 μm by 500 μm.

A human red blood cell is only 8 μm in diameter, equivalent to 0·008 mm.

The table below shows the relative sizes of some structures.

Cell/cell structure	Approximate length/diameter μm
Virus	1
Bacterial cell	3
Mitochondrion	4
Cheek cell nucleus	5
Red blood cell	8
Sperm	60
Egg cell	130

TOP TIP

You must be able to move between these different units so make sure you practise converting one to the other and have an awareness of the relative sizes of cells.

Quick Test

1. Name the material of which plant cells, walls are composed.

2. Name the part of the cell that carries out aerobic respiration.

3. Name the cell structures common to both plant and yeast cells.

Transport across cell membranes

Cell membrane

The cell membrane is the structure across which substances must pass if they are to enter or leave a cell. However, the membrane is selective about what can/can't pass across; it is termed **selectively permeable**. Small molecules, such as carbon dioxide, oxygen and water, can pass across easily but larger molecules, such as proteins, cannot.

The structure of the membrane is responsible for these properties and it consists of a layer of two different molecules: proteins and **phospholipids** arranged as shown below:

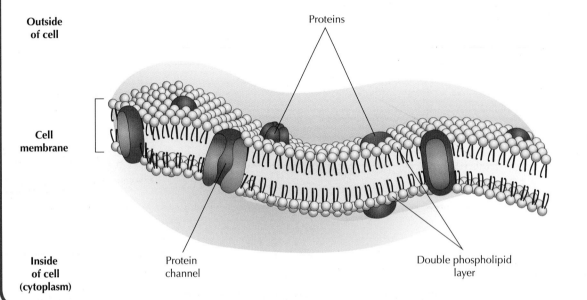

Outside of cell

Proteins

Cell membrane

Inside of cell (cytoplasm)

Protein channel

Double phospholipid layer

Passive transport and concentration gradient

Passive transport occurs when molecules of a substance move down a **concentration gradient**. This does not require energy. An example of this is the process of **diffusion**. When molecules of gases, liquids or other dissolved substances move they pass down a concentration gradient from a higher to a lower concentration. The net movement of these molecules will only stop when its concentration is equal.

Diffusion allows living cells to move many different substances in and out of cells and is a

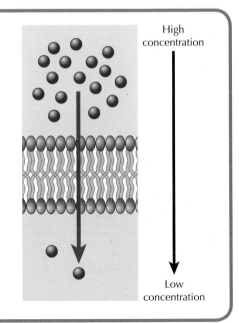

High concentration

Low concentration

vital means for the exchange of important substances between the inside and outside of cells. Here are some examples in an animal cell:

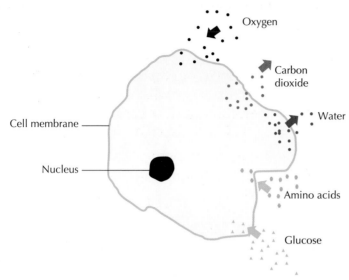

The actual 'direction' of the concentration gradient is dependent on a number of factors. For example, in the lungs the oxygen gradient is from high oxygen concentration outside the cell to low oxygen concentration inside. This gradient will be reversed for cells elsewhere in the body.

TOP TIP

Be able to give examples of diffusion in and out of animal and plant cells in different situations.

Osmosis

The process of **osmosis** is also an example of passive transport. Osmosis is simply the diffusion of water molecules. This involves the movement of water molecules from an area of high water concentration to an area of lower water concentration across a selectively permeable membrane.

In the example below, the sugar molecules do not move so easily or quickly as the water molecules and so there is a change in the volume of solution on either side of the selectively permeable membrane after a time.

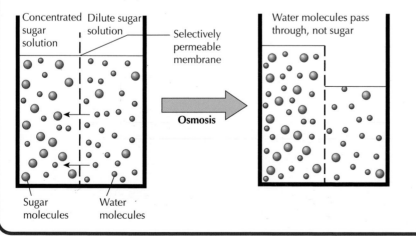

TOP TIP

Remember that osmosis is a special case of diffusion involving the movement of water across a selectively permeable membrane.

The movement of water by osmosis is important for both plant and animal cells. For example, freshwater fish constantly take in water by osmosis across their gills, which act like selectively permeable membranes. In single-celled organisms, water moves across their cell membranes by osmosis. Cells in an animal's body behave similarly.

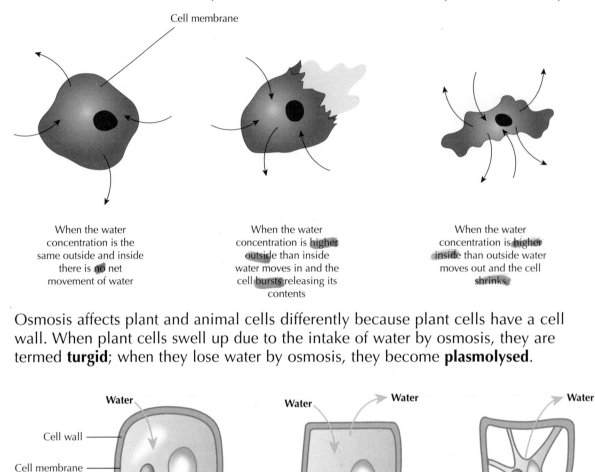

Cell membrane

When the water concentration is the same outside and inside there is no net movement of water

When the water concentration is higher outside than inside water moves in and the cell bursts releasing its contents

When the water concentration is higher inside than outside water moves out and the cell shrinks

Osmosis affects plant and animal cells differently because plant cells have a cell wall. When plant cells swell up due to the intake of water by osmosis, they are termed **turgid**; when they lose water by osmosis, they become **plasmolysed**.

Water

Cell wall

Cell membrane

Vacuole

High water concentration outside cell and water moves in by osmosis causing cell to become turgid

Water

Water concentrations inside and outside are the same so no net movement of water

Water

Higher water concentration inside the cell so water moves out by osmosis causing the cell be become plasmolysed

Active transport

Sometimes diffusion and osmosis are not sufficiently quick or effective for moving substances in and out of cells. For example, the cells in plant roots need to take in nutrients, such as mineral ions, from the soil, where they are in low concentration, into the cells, where they are usually in high concentration. This is therefore against the concentration gradient. Similarly, glucose is moved out of the gut into the bloodstream against a concentration gradient. This type of movement is energy-demanding and is therefore called **active transport**.

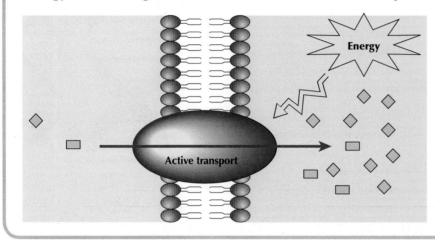

Quick Test

1.   Explain the cell membrane described as 'selectively permeable'?
2.   Name three substances that can enter a cell by diffusion.
3.   Give two ways in which active transport differs from diffusion.

DNA and production of proteins

Structure of DNA

The chromosomes in the nucleus of living cells contain a chemical called **deoxyribonucleic acid**, commonly shortened to **DNA**. Each molecule of DNA is made up of two strands twisted around each other to form a coiled, ladder-like shape called a **double helix**.

Each strand is, in turn, composed of many repeating building blocks. A DNA nucleotide consists of a 5-carbon sugar called **deoxyribose**, a phosphate grouping and one of four nitrogen-containing chemicals, **adenine**, **thymine**, **guanine** or **cytosine**, collectively called **bases**.

Each of the two strands is held together by weak links between the bases, which are in the middle of the double helix. The phosphate and deoxyribose form the backbone of the double helix. The bases are linked together in **complementary** pairings.

Adenine always pairs with thymine, while cytosine always pairs with guanine. The sequence of the bases is responsible for the types of protein made in cells.

A length of DNA which codes for a specific sequence of amino acids to make a protein is called a **gene**.

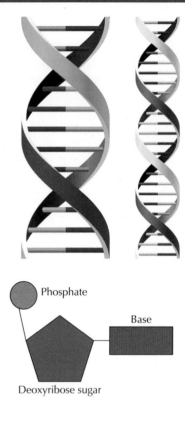

Phosphate

Base

Deoxyribose sugar

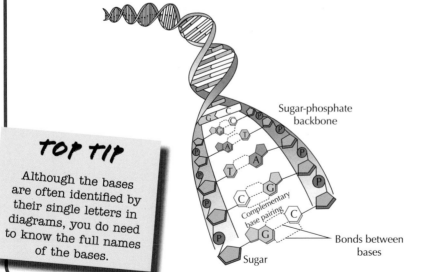

Sugar-phosphate backbone

Complementary base pairing

Bonds between bases

Sugar

TOP TIP

You will study genes in more detail as you move through the course.

TOP TIP

Although the bases are often identified by their single letters in diagrams, you do need to know the full names of the bases.

TOP TIP

The differences that exist between living things are due to the differences in the sequence of the bases in their nucleic acids.

DNA and protein

All the reactions that go on inside a cell are controlled by **enzymes**. Enzymes are made of proteins built up from units called **amino acids**.

—Amino acids—

⬇

Join up in particular order to form protein

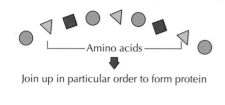

Enzymes act as biological **catalysts** and are responsible for the function, growth and development of cells and whole organisms. Enzyme production is controlled by DNA. The actual structures that join up the amino acids in a particular order, determined by the base sequence of the DNA, to form proteins in cells, are the ribosomes found in the cytoplasm. The ribosomes and DNA are 'linked' by a special carrier molecule called **messenger ribonucleic acid** or **mRNA** for short. Messenger RNA carries a complementary copy of the genetic code from the DNA.

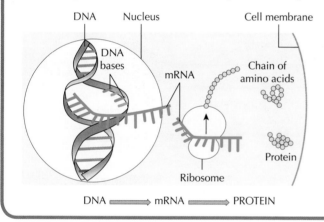

Quick Test

1. Name the three types of chemicals needed to build the DNA molecule.
2. Give the correct complementary pairing of adenine (A), thymine (T), cytosine (C) and guanine (G).
3. Explain the role of mRNA.

Proteins

Introduction

Many important chemicals are made up of proteins. The variety of structures and functions of proteins arise from the different sequence of the amino acids present.

Protein functions	Description
Enzymes	Catalyse chemical reactions in a cell
Hormones	Chemicals that act as messengers and affect how living things develop and behave
Antibodies	Large molecules that help defend against infections
Haemoglobin	Transports oxygen in the bloodstream

Proteins that form parts of a cell, or give it support and rigidity, are termed **structural**. Some proteins can act as **receptors**. A receptor is a group of molecules, usually in a cell membrane, which fit another complementary molecule so that when these link a change in cell function is brought about.

Enzymes

TOP TIP

Remember, it is the sequence of the bases in the DNA that determines the sequence of the amino acids in a protein. The protein will then have a specific structure and function.

Enzymes function as biological catalysts and are made by all living cells. They speed up cellular processes and are unchanged in the process. Some enzymes catalyse **degradation** reactions, where large molecules are broken down. Others catalyse **synthesis** reactions, where smaller molecules are built up into larger molecules. Each enzyme is specific for the chemical on which it acts, called the **substrate**. This means that an enzyme will react with only one particular substrate. The resultant chemicals are called the products. The area where the enzyme and substrate meet is called the **active site**, which is complementary in shape to the specific substrate.

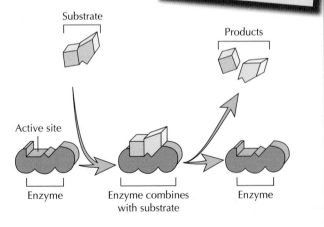

Substrate

Products

Active site

Enzyme

Enzyme combines with substrate

Enzyme

Factors affecting enzyme activity

Two important factors that affect enzyme activity are pH and temperature. This is because enzymes are made of proteins, which are sensitive to changes in the pH and temperature of their surroundings. Each enzyme has a particular pH and temperature at which it is most active called the **optimum**.

Most are most active at a pH near neutral (7) but others, such as those in the stomach, work best in acidic conditions; those found in the liver and bone work best in alkaline conditions.

Enzymes are also very sensitive to changes in temperature, working best at an optimum determined by the type of organism in which the enzyme-catalysed reaction is taking place. In humans with a body temperature of about 37°C, the optimum temperature for enzyme function is also very close to 37°C.

When a factor, such as pH or temperature, goes too much above or below the optimum, enzyme function slows down. If the change in structure (and hence function) is irreversible, the enzyme is said to be **denatured** and the active site is no longer able to bind with its specific substrate.

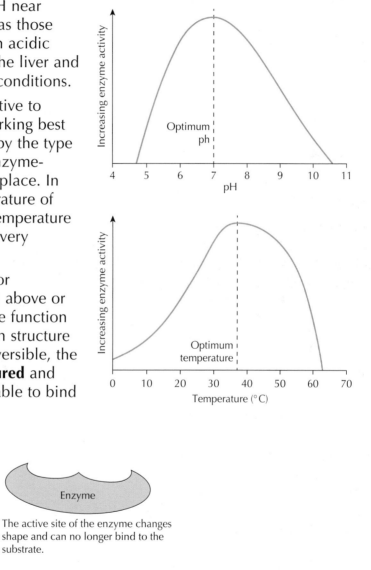

The active site of the enzyme changes shape and can no longer bind to the substrate.

Quick Test

1. Explain why the sequence of amino acids in a protein so important.
2. Explain why one enzyme that breaks down starch is not effective in breaking down protein.
3. Suggest one possible explanation why storing food in a fridge extends its shelflife.

Genetic engineering

Genetic engineering

It is now possible artificially to transfer specific parts of a cell's genetic material, called **genes**, from one **species** to an entirely different species. These techniques come under the general heading of **genetic engineering**. This process enables the **genetically modified (GM)** cell or organism to make or do something it did not before. For example, the following sequence is used to genetically modify a bacterial cell to make the human hormone **insulin**.

Gene on human chromosome for producing
human insulin is identified

Insulin producing gene is extracted

Bacterial plasmid is extracted and opened

Gene is inserted into the bacterial plasmid

Plasmid is put back into host the bacterial cell

Bacterial cell allowed to reproduce

Multiple copies of the plasmids are made

Bacteria produce insulin, which is harvested

> **TOP TIP**
>
> Make sure you know the sequence involved in genetically modifying a cell and the function of a plasmid as a carrier of genes.

> **TOP TIP**
>
> You need to be able to understand information presented in different ways, sometimes in text form, sometimes a diagram etc.

Notice that the bacterial plasmid is acting as a 'carrier' to transport the human gene into the bacterial cell.

Enzymes are involved in the cutting open of plasmids, removal of gene from chromosome and sealing the gene into the plasmid.

The diagram below shows the same process.

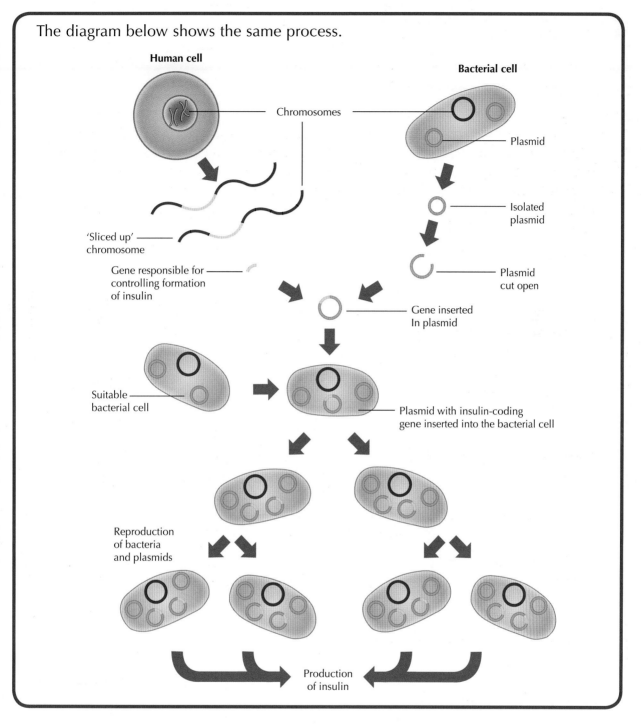

Quick Test

1. Describe the structure and function of a gene.

2. Suggest two reasons why people might be concerned about the application of genetic engineering to produce genetically modified (GM) foods.

3. State an example of a carrier found in bacterial cells which can be used in genetic engineering.

Respiration

Introduction

Glucose is an energy-rich molecule but this stored energy has to be released in a controlled way to prevent damage to cells. To ensure the safe release of this energy, a large number of small steps are used. Each step is under the control of an enzyme; collectively, these steps are called **respiration**.

ATP and respiration

TOP TIP

ATP is not an energy store but transfers energy.

The energy released from glucose is not used directly by living cells but is used to generate ATP. ATP is a chemical universally found in plant and animal cells as the source of immediate energy.

This ATP can be broken down directly to drive any reaction that needs energy and then be reformed as needed using energy from the breakdown of more glucose.

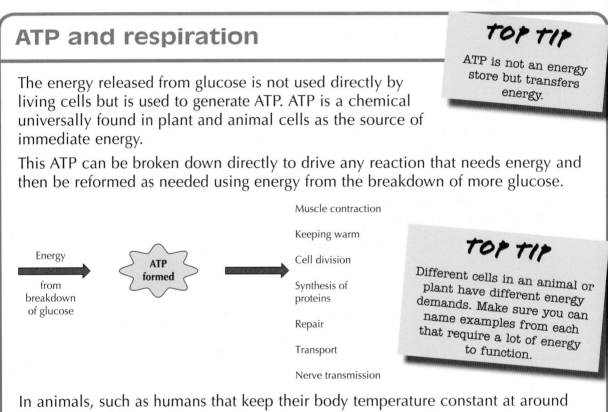

Muscle contraction

Keeping warm

Cell division

Synthesis of proteins

Repair

Transport

Nerve transmission

TOP TIP

Different cells in an animal or plant have different energy demands. Make sure you can name examples from each that require a lot of energy to function.

In animals, such as humans that keep their body temperature constant at around 37°C, the heat is supplied by cells as they respire glucose. It is no surprise their enzymes are more active at this temperature!

Aerobic respiration

The breakdown of glucose normally takes place in the presence of oxygen and for this reason is called **aerobic**. This form of respiration results in the complete breakdown of the glucose to form carbon dioxide and water as well as ATP.

This process is a series of steps, each catalysed by its own specific enzyme. The first stage of aerobic respiration takes place in the cell cytoplasm and results in the production of two molecules of **pyruvate** from the breakdown of a glucose molecule. Two molecules of ATP are generated. In the presence of oxygen the pyruvate is then further broken down in the mitochondria of the cell to produce carbon dioxide, water and ATP. Since the glucose is completely broken down in aerobic respiration, this is a very efficient process. Aerobic respiration yields a large number of ATP molecules.

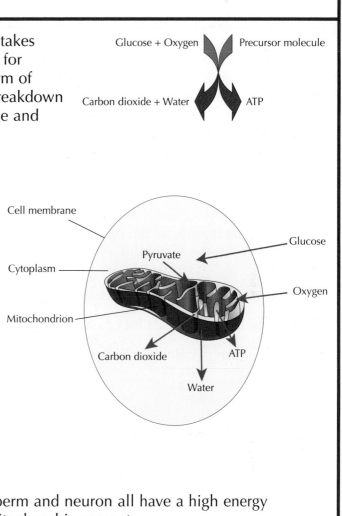

Cells such as muscle, companion, sperm and neuron all have a high energy demand and therefore have many mitochondria present.

Respiration without oxygen in animals

Sometimes, for short periods only, animals can respire without oxygen. This type of respiration uses a **fermentation** pathway. For example, during strenuous exercise, it might not be possible to meet the excessive oxygen demands of working muscles, which use up all the available oxygen. In this situation, glucose cannot be completely respired. The pyruvate is instead converted to **lactate**.

If lactate accumulates in the bloodstream, it causes cramp and stops the muscles working efficiently. Anaerobic respiration is not a very efficient process for several reasons:

1. Glucose is only partly broken down

2. Lactate still contains a lot of energy

3. Only two molecules of ATP are produced.

TOP TIP

You need to know why the fermentation pathway is much less efficient than aerobic respiration.

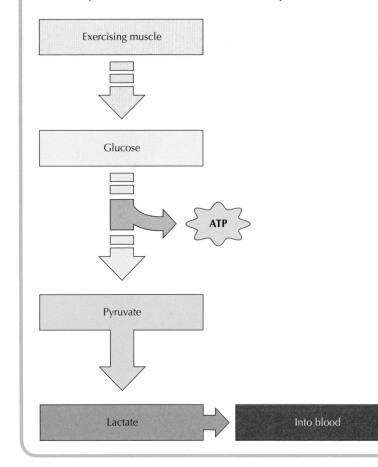

Respiration without oxygen in plant and fungal cells

In some situations, cells can respire without oxygen for periods of time, but not indefinitely. For example, if plant roots get flooded with water and deprived of oxygen, they will switch to using the fermentation pathway until oxygen becomes available. The pyruvate is converted into an alcohol called **ethanol** and carbon dioxide.

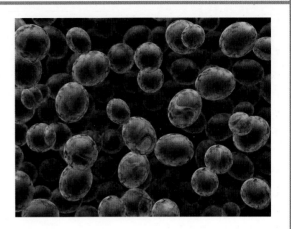

Yeast cells can break down glucose in the absence of oxygen by switching to the fermentation pathway.

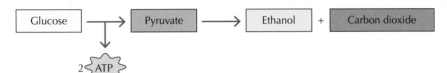

Aerobic respiration	Anaerobic respiration/fermentation
Needs oxygen	Does not need oxygen
Efficient in producing 38 ATP molecules for every molecule of glucose respired	Inefficient in producing only two ATP molecules for every molecule of glucose respired
Can carry on indefinitely	Can only last relatively short periods of time
Glucose completely broken down	Glucose not completely broken down
Endproducts are carbon dioxide and water	Endproducts are lactate in animal cells, and ethanol and carbon dioxide in plant and yeast cells
Starts in the cytoplasm and is completed in the mitochondria	Takes place only in the cytoplasm

Quick Test

1. Explain why respiration takes place in many small steps.

2. Name the common simple sugar that is used as a substrate for respiration.

3. Copy and complete the following word equation for aerobic respiration:

_____ + oxygen ⟶ carbon dioxide + _____ + _____

Revision Questions

Section A

1. The diagram below shows a human cell viewed under a high-powered microscope.

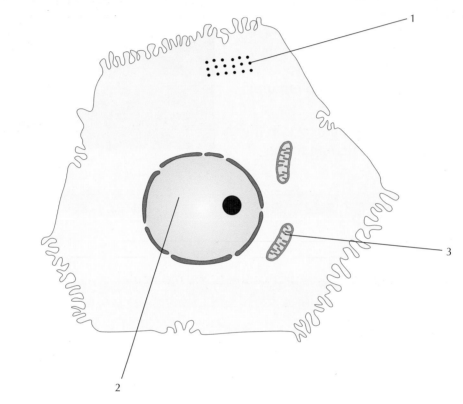

   Which of the following correctly identifies structures 1, 2 and 3?

	1	2	3
A	Mitochondrion	Cell membrane	Nucleus
B	Cell membrane	Mitochondrion	Ribosomes
C	Nucleus	Cell membrane	Ribosomes
D	Ribosomes	Nucleus	Mitochondrion

2. If a mitochondrion appears to be 12 mm under a magnification of × 12 000, what is its size in µm?

   A 0·5

   B 1·0

   C 1·5

   D 2·0

3. A piece of plant material was weighed and then placed in strong sugar solution for 45 minutes then in pure water for 45 minutes. It was removed, dried and reweighed after each immersion.

   Which of the following is very likely to be the readings obtained?

	Initial mass (g)	After 45 mins in strong sugar (g)	After 45 mins in water (g)
A	10	12	10
B	8	7	6
C	12	10	14
D	9	8	7

4. When a yeast cell breaks down one molecule of glucose in the absence of oxygen, how many ATP molecules are gained?

   A   1

   B   2

   C   36

   D   38

5. The following diagram shows part of the DNA molecule.

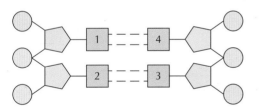

   If 4 and 3 represent adenine and cytosine respectively, which of the following correctly represents 1 and 2?

	Thymine	Guanine
A	2	2
B	1	1
C	1	2
D	2	1

Section B

1. The following graph shows the change in the rate of an enzyme-catalysed reaction with increasing temperature. The rate is measured in 'units'.

    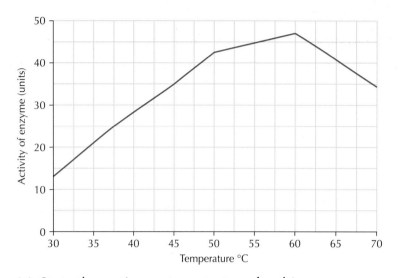

    (a) State the optimum temperature for this enzyme. [1]

    (b) State the percentage increase in the activity of the enzyme when the temperature is increased from 37·5°C to 45°C. [1]

    (c) (i) State if this enzyme likely to be from an animal. [1]

    (ii) Explain your answer. [2]

2. (a) Copy and complete the table which compares aerobic respiration with fermentation in a yeast cell using the letters of the descriptions given. [2]

    A oxygen is not required

    B most ATP generated from one glucose molecule

    C oxygen is required

    D ethanol is produced

Aerobic respiration	Fermentation

    (b) Name one substance produced by a cell carrying out aerobic respiration. [1]

3. (a) The diagram below shows a plant cell after immersion in three different solutions.

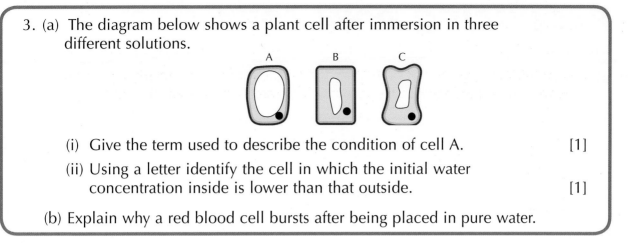

(i) Give the term used to describe the condition of cell A. [1]

(ii) Using a letter identify the cell in which the initial water concentration inside is lower than that outside. [1]

(b) Explain why a red blood cell bursts after being placed in pure water.

Producing new cells

Introduction

TOP TIP

Make sure you understand why mitosis must maintain the number of chromosomes in each daughter cell.

Whenever an organism grows, or if repair or replacement of damaged or worn-out cells is required, new cells are needed. This process is under the control of the cell nucleus. Since the cell's genetic instructions are stored in the nucleus, it is essential that the process of forming new 'daughter' cells preserves exactly, with no loss or gain, this genetic material, which is in the form of structures called **chromosomes**. The division of the cell nucleus is called **mitosis** and the process ensures that the new cells each have the same **chromosome complement** as the original 'parent' cell. Any change to the chromosome complement could have a serious impact on the proper working of a cell.

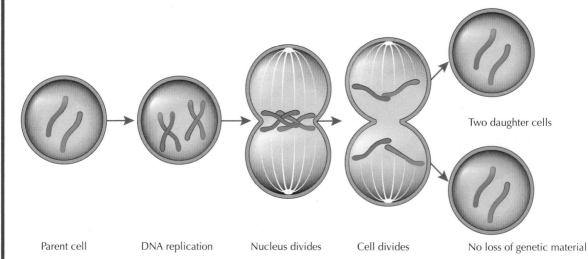

Parent cell   DNA replication   Nucleus divides   Cell divides   No loss of genetic material

Two daughter cells

The chromosomes in a cell are not usually visible until the cell is about to divide by mitosis, when they become short and thick and easily seen under the microscope. Each chromosome is then apparent as two identical strands called **chromatids**. During mitosis, fibres which make up a **spindle**, attach to these centromeres on the chromosomes. The chromosomes are then orientated along the central plane of the cell called the **equator**.

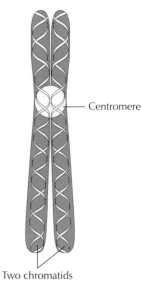

Centromere

Two chromatids

got it? ☐ ☐ ☐

Process of mitosis

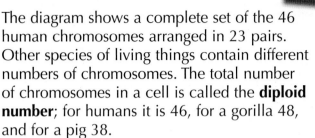

The diagram shows a complete set of the 46 human chromosomes arranged in 23 pairs. Other species of living things contain different numbers of chromosomes. The total number of chromosomes in a cell is called the **diploid number**; for humans it is 46, for a gorilla 48, and for a pig 38.

Mitosis ensures that every daughter cell gets exactly the same chromosomes as its parent cell. While this takes place continuously, it is easier to study if considered in stages.

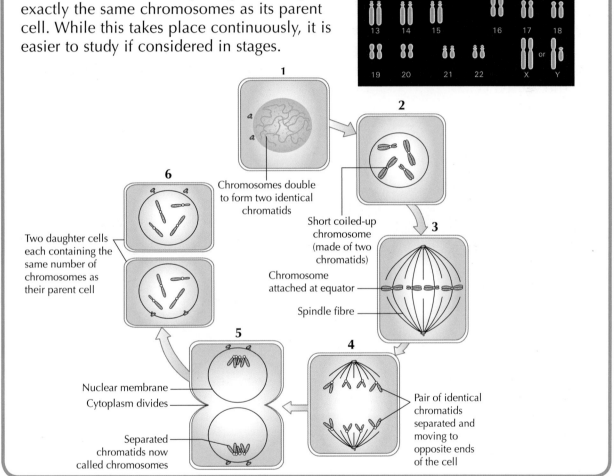

1
Chromosomes double to form two identical chromatids

2
Short coiled-up chromosome (made of two chromatids)

Chromosome attached at equator

Spindle fibre

3

4
Pair of identical chromatids separated and moving to opposite ends of the cell

5
Nuclear membrane
Cytoplasm divides
Separated chromatids now called chromosomes

6
Two daughter cells each containing the same number of chromosomes as their parent cell

Quick Test

1. The process of mitosis ensures that the daughter cells all receive the same diploid number of chromosomes. Explain why this is important.

2. State the diploid number of the cell shown in the diagram.

3. Name the structure which pulls the chromatids apart during mitosis.

Stem cells in animals

It has been estimated that in one day, you will shed about a million skin cells! These have to be constantly replaced. This requires a supply of new cells as well as maintaining the original population of cells which gives rise to the new skin cells. The original population is made up of cells that are not yet specialised but can form skin cells. These unspecialised cells are called **stem cells** and are extremely important in animals. They have three important features:

1. They are not yet specialised

2. They can divide repeatedly without limit during an animal's lifetime

3. They divide to form two daughter cells, each of which may remain a stem cell or become another type of specialised cell.

TOP TIP

Stem cells are found in many different places in an animal's body such as bone marrow, teeth, lungs, heart and brain.

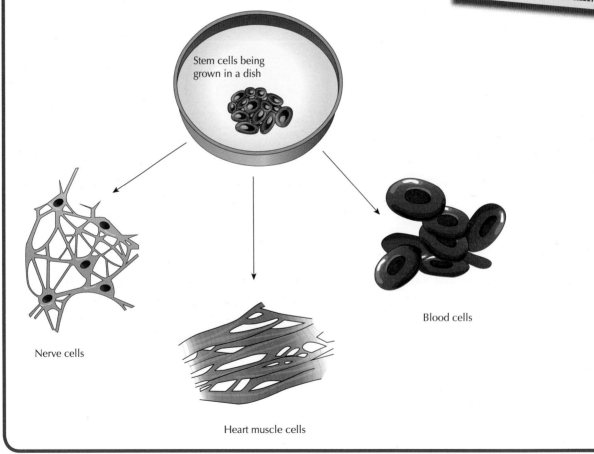

Stem cells being grown in a dish

Nerve cells

Heart muscle cells

Blood cells

Adult stem cells function to repair and replace damaged or worn-out cells in the tissues where they are found. Research into the uses of stem cells involves looking at ways to grow them in large numbers, then making them produce specific types of cells to treat injuries or diseases. Their use is not without some ethical issues, particularly in relation to the source of the stem cells.

Specialisation of cells

While cells of different living things share many features in common, they show great variation in their size, shape and function. In more complex, **multicellular** organisms, cells of similar size, shape and function often group together to form **tissues**.

Cells that form tissues are usually **specialised** so that different jobs can be done more effectively. Tissues, in turn, are grouped to form **organs**. Organs in turn form **systems** which collectively form the whole organism.

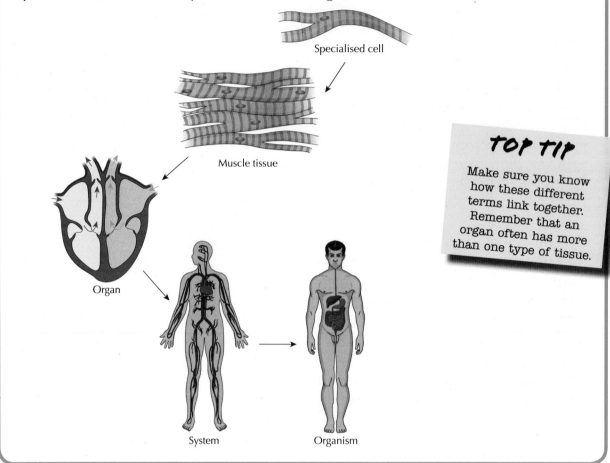

Specialised cell

Muscle tissue

Organ

System

Organism

TOP TIP

Make sure you know how these different terms link together. Remember that an organ often has more than one type of tissue.

Quick Test

1. State three features of a stem cell.

2. Which of the following statements is NOT true?

   a) Stem cells cannot grow in large numbers.

   b) Growth takes place in restricted areas of an animal.

   c) Stem cells are not committed to becoming a particular type of cell.

   d) During the lifetime of an animal, stem cells can divide repeatedly.

3. State the term used to describe an organism made up of many cells.

4. State the general term which describes a group of cells that are all the same size and shape and perform a similar function.

5. The leaf of a plant carries out photosynthesis. It consists of groups of cells that trap sunlight energy, transport water, allow gas exchange and protect the leaf. Give the general term which describes a leaf as a collection of these different groups.

Control and communication

Structure and function of the central nervous system

To preserve and maintain life, a multicellular animal requires an efficient means of detecting changes in its environment such as sound, temperature increase or decrease, light intensity, etc. Such changes are called stimuli. The animal must then be able to react to those stimuli in an appropriate way.

Additionally, the animal must be able to monitor its 'internal environment' to control the rates of heartbeat and breathing, and check for constancy in the pH of the blood, levels of nutrients, etc. The **nervous system** is responsible for this control.

All over an animal's body there are special cells or groups of cells that can detect specific stimuli. These **sensory receptors** send information about what is happening externally as well as internally to the **central nervous system (CNS)**. The central nervous system consists of the **brain** and **spinal cord**. The spinal cord is protected by a series of bones that make up a tube called the **spinal column**.

The main functions of the nervous system in an animal's body are to:

- send information to all parts
- receive information from all parts
- make sense of the information received
- co-ordinate the workings of internal organs
- respond to stimuli in the external environment.

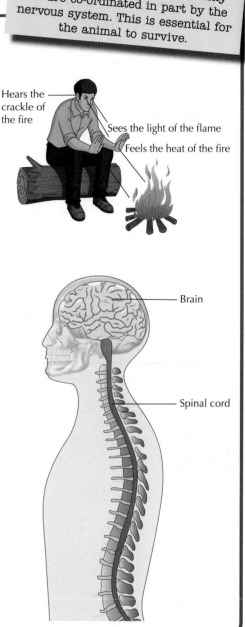

Hears the crackle of the fire

Sees the light of the flame

Feels the heat of the fire

Brain

Spinal cord

The brain

This remarkable organ consists of three important structures: **cerebrum** (divided into two halves called the **cerebral hemispheres**), **cerebellum** and **medulla**. Each performs particular functions.

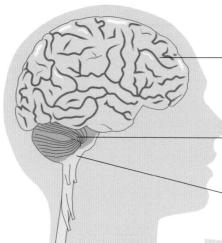

Cerebrum controls all higher activities such as memory, emotions, sensations, conscious decisions, processing information and intelligence

Cerebellum controls balance and co-ordinates muscles involved in precise and accurate movements

Medulla controls activities that the animal is not consciously aware of, such as breathing, digestion and heartbeat

TOP TIP

Make sure you can label a diagram of the brain and know what the functions of each of the three important parts are.

Reflex action

Multicellular animals inherit the ability to respond quickly and without thinking to some stimuli, particularly ones that might be harmful. For example, the simple act of a dog blinking its eyes is an unconscious mechanism for constantly cleaning and keeping the surface of the eye moist and thereby inhibiting infection. Such unconscious actions are called **reflexes**. Sneezing, coughing, blinking, withdrawing from a hot or sharp object are all examples of reflexes. In animals with short lifespans, reflexes save valuable time and energy as they don't need to be learned and help the animal survive because it can focus on other immediate aspects of its environment.

TOP TIP

There are many examples of reflexes that help an animal survive. Do some research to find some examples of these and how they increase the chances of survival.

In general, reflexes:

- give rise to a 'pre-programmed' type of response to a given stimulus
- are usually not under conscious control although, sometimes, they can be influenced by a voluntary input

- give rise to very rapid responses to appropriate stimuli
- generally cannot be stopped once they have started
- often do not involve the brain in any way
- usually follow very simple pathways involving few nerve cells.

Reflex arc

A **reflex arc** is a very simple pathway for a reflex action and typically involves three special nerve cells, sometimes called **neurons**, acting in series. At either end is a **receptor**, which detects a specific **stimulus** and an **effector**, which performs some kind of action. The three neurons involved are:

1. **sensory neuron** carrying information from a receptor to an inter neuron

2. **inter neuron** connects a sensory neuron to a motor neuron

3. **motor neuron** carrying information towards an effector which may be a muscle or **gland.**

> **TOP TIP**
>
> Remember neurons don't physically touch each other. They are separated by synapses.

Between one neuron and the next is a tiny gap called a **synapse**. When a nerve impulse reaches the end of one neurone, a small quantity of a chemical is released, which diffuses across the gap to the next neuron. This is then stimulated to carry the impulse, rather like a baton being passed from one runner to the next in a relay race.

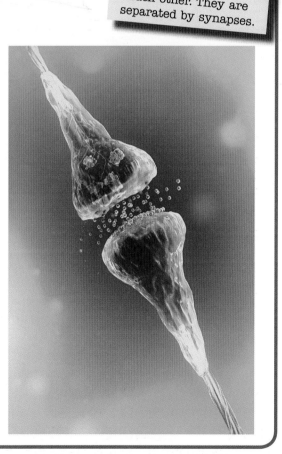

Synapses allow:

- nerve impulses to travel in one direction because the chemical is released from one neuron and travels across the gap to the next neuron but not the other way around
- multiple connections between different neurons to be made
- control of the ongoing nerve impulse by altering the type and quantity of chemical released so that the ongoing impulse may be strengthened or weakened or prevented entirely.

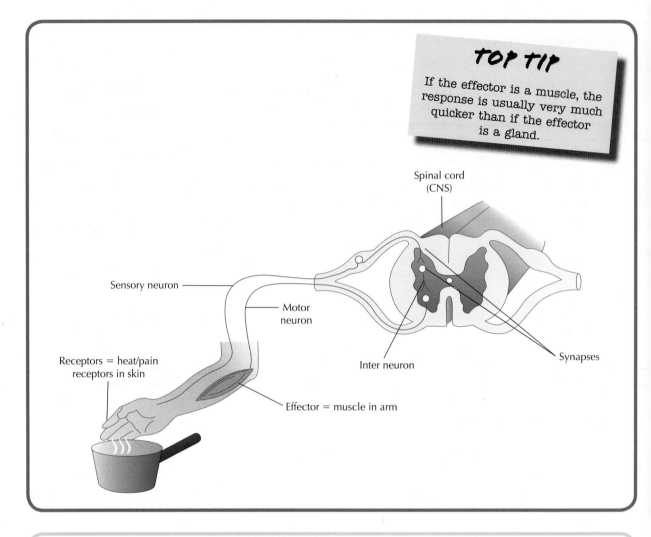

TOP TIP

If the effector is a muscle, the response is usually very much quicker than if the effector is a gland.

Spinal cord (CNS)

Sensory neuron

Motor neuron

Synapses

Receptors = heat/pain receptors in skin

Inter neuron

Effector = muscle in arm

Hormonal control

Sometimes a slower response is required to a stimulus, or the cells affected by that stimulus are located in different parts of an animal's body. In these situations, the **endocrine system** for co-ordination comes into operation. This contains many **endocrine glands**, which release **hormones** directly into the bloodstream. Hormones are chemical messengers. A **target tissue** has cells with complementary receptor proteins for specific hormones, so only that tissue will be affected by these hormones.

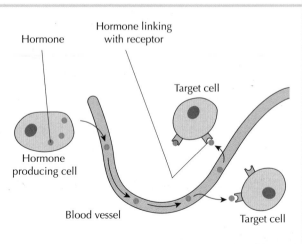

Hormone

Hormone linking with receptor

Target cell

Hormone producing cell

Blood vessel

Target cell

Blood glucose regulation

Glucose is one of the most important sources of energy for most animals. It is vital that the levels of glucose in the bloodstream don't go above or below a critical value or the proper working of body cells is reduced. If this occurs and continues, eventually the whole animal will be affected.

> **TOP TIP**
>
> Glucose is the only source of energy for brain cells.

The source of an animal's glucose is its food but since animals don't eat all the time, they must store glucose until it is needed. After eating, the level of glucose is very high and so the excess is stored in the form of a chemical called **glycogen** in the liver. Glycogen has two important properties:

1. Unlike glucose, which is soluble, glycogen is insoluble and so can be easily stored

2. Because it is insoluble, glycogen does not cause osmosis to take place as glucose does.

To promote the conversion of glucose to glycogen, the hormone insulin is produced by an endocrine gland called the **pancreas**. When glucose is required, the pancreas is able to produce another hormone called **glucagon** which converts the glycogen back into glucose.

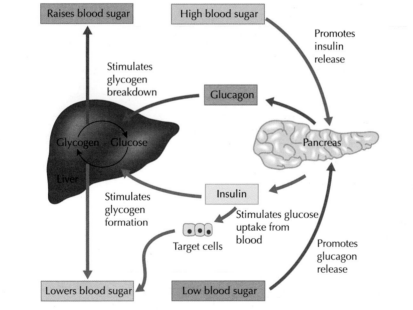

In this way, the level of glucose in the blood is kept at a constant level.

Quick Test

1.	Name the two body systems responsible for control and communication.
2.	Explain what a 'stimulus' is.
3.	Explain why glycogen is such a good storage carbohydrate.

Reproduction

Diploid and haploid cells

The cells that make up the body of a plant or animal contain the **diploid number** of chromosomes (n). The diploid number is made up of two sets of chromosomes. For example, in a pigeon it is 80 and in rice it is 24. During reproduction, special **sex cells** or **gametes** are formed, which contain half the diploid number. This is called the **haploid number** (n).

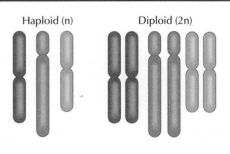

Haploid (n)   Diploid (2n)

Gametes

In animals, sex cells are produced in structures called **gonads**. These are usually the **testes** (singular **testis**), which produce **sperm**, and **ovaries**, which produce **eggs** or **ova** (singular **ovum**).

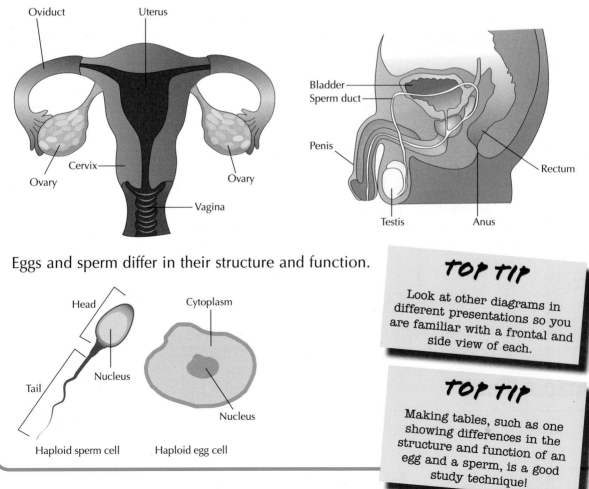

Eggs and sperm differ in their structure and function.

Haploid sperm cell   Haploid egg cell

TOP TIP

Look at other diagrams in different presentations so you are familiar with a frontal and side view of each.

TOP TIP

Making tables, such as one showing differences in the structure and function of an egg and a sperm, is a good study technique!

In flowering plants, the female gametes are within the **ovules** and formed in the ovary. The male gametes are within the **pollen grains** and formed in the **anther**.

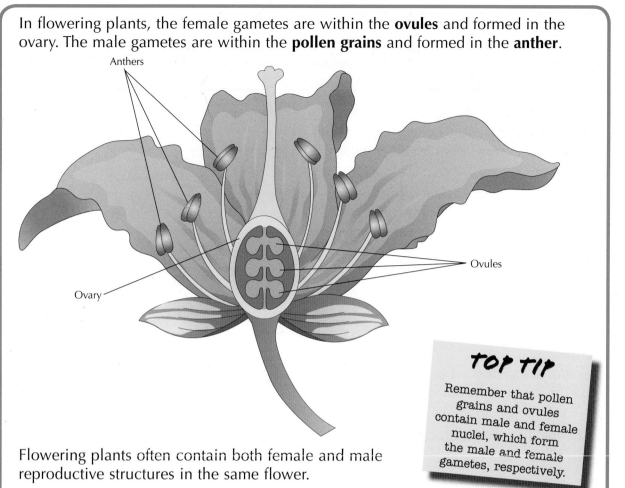

Flowering plants often contain both female and male reproductive structures in the same flower.

TOP TIP

Remember that pollen grains and ovules contain male and female nuclei, which form the male and female gametes, respectively.

Gametes in both animals and plants each:

- contain only half the diploid number of chromosomes and therefore only half the genetic material of normal body cells
- are unique combinations of genetic material from the parent organism that produced them.

The way in which gametes are produced in plants and animals allows for unlimited variation in the new organisms formed in subsequent generations. Each parent contributes, randomly, half its genetic material to form a gamete so that the newly formed organism is a unique combination of genes from each parent, giving rise to the variety of different forms of a species.

Fertilisation

For a new individual to form, a male and female gamete must fuse to form a **zygote,** which divides to form an **embryo**. This event is called **fertilisation**. In animals, fertilisation involves sperms and eggs.

Fertilisation restores the diploid number of chromosomes by combining two haploid gametes. For example, in humans, each gamete carries 23 chromosomes (n).

The life cycle of a multicellular organism alternates between a diploid state for the body cells and a haploid state for the gametes.

Sperm are deposited inside the vagina by the erect penis during sexual intercourse.

The fusion event takes place in the **oviduct**.

Ovum (egg) with its 23 chromosomes

Sperm with its 23 chromosomes

TOP TIP

Remember that only one sperm fertilises an ovum.

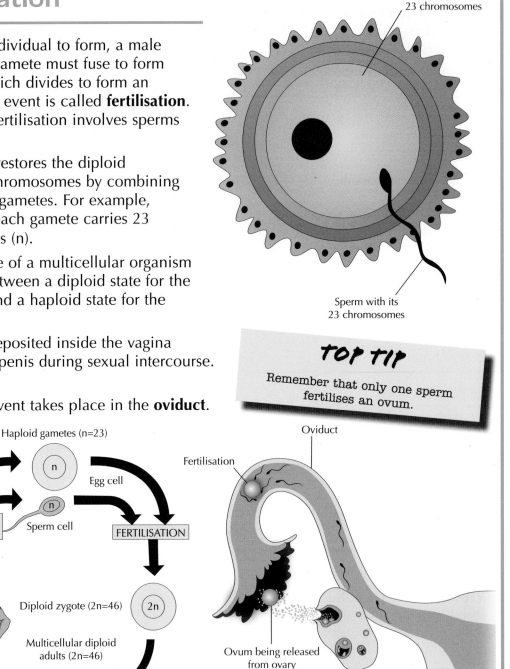

Haploid gametes (n=23)

n

Egg cell

n

FORMATION OF GAMETES

Sperm cell

FERTILISATION

Diploid zygote (2n=46)

2n

Multicellular diploid adults (2n=46)

Mitosis and development

Oviduct

Fertilisation

Ovum being released from ovary

In plants, the male gamete inside the pollen grain fuses with the female gamete inside the ovule. To reach the ovule, the pollen grain first has to grow a **pollen tube** down which the male gamete travels. The fertilisation event takes place inside the ovule to form a zygote.

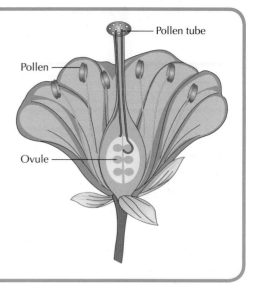

Pollen tube

Pollen

Ovule

TOP TIP

Don't confuse fertilisation (fusion of male and female gametes) with pollination (transfer of pollen) in plants.

Quick Test

1. State two ways in which the structure of a sperm and the structure of an egg differ.

2. If a cell in the leaf of a flowering plant contains eight chromosomes, calculate how many chromosomes would be found in an ovule.

3. Explain the significance of gametes carrying unique combinations of genetic material from the parent organism which produced them.

Variation and inheritance

Variation

Living things differ from species to species and within the same species. Such differences are called **variation**. Variation takes two different forms, **discrete** (sometimes called **discontinuous**) variation and **continuous variation**.

Discrete variations include eye colour, right- and left-handedness and flower colour, while continuous variations include weight, height, rate of heartbeat, length of leaves and number of petals.

Phenotype

The appearance of a plant or animal is a direct result of the genes that are inherited from the parents. The random formation of this combination of genes gives rise to variation, while the actual expression of the genes, often the 'outward appearance' of an individual, is called the **phenotype**. The phenotype may not always be 'outward', for example, in the case of a person's blood grouping.

Some single gene features such as tongue-rolling ability and the presence or absence of ear lobes in humans, plant height in rice, and shape of fruit in sweet peppers are controlled by only one gene.

Most features of an individual's phenotype are not monogenically controlled but are under the influence of many genes. This is called **polygenic inheritance**, which is often affected by the environment.

Polygenic inheritance normally shows continuous variation and typically shows a distribution that is termed 'normal'. Height is a well understood example of polygenic inheritance. People are not just short or tall but show variation in heights. Height is influenced by the environment. A person born with the genes to become tall may indeed not fulfil their potential because of a lack of food or disease of some kind.

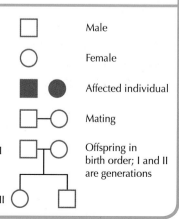

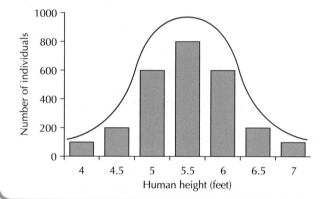

> ## TOP TIP
> Skin colour and weight are also well-known examples of polygenic inheritance.

> ## TOP TIP
> Remember this distinction: single gene measurements fall into distinct groups, but polygenic measurements have a range of values between a minimum and a maximum.

Family trees

Using **family trees** is one way of recording patterns in inheritance. They are often used to help people with a history of genetic diseases, but are used widely in animal and plant genetics. Typically, in animal family trees, the same symbols are commonly used.

☐ Male

○ Female

■ ● Affected individual

☐—○ Mating

I ☐—○ Offspring in birth order; I and II are generations

II ○ ☐

Albinism is a condition in humans that results in a lack of skin pigmentation. Family trees are often used to give advice to people of risks to themselves or their children; born, or unborn.

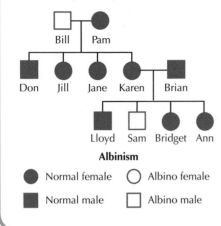

Albinism

● Normal female   ○ Albino female

■ Normal male   □ Albino male

Inheritance

How characteristics are passed on from one generation to another is called **inheritance**. Such characteristics are controlled by genes, which exist in pairs and are found on the chromosomes. Since chromosomes are normally inherited in equal numbers from each parent, one member of each pair of genes comes from the male and one from the female parent. Genes can exist in different forms called **alleles** usually represented by letters. Different alleles of a gene can be **dominant** or **recessive**. A **homozygous** individual has two alleles of a gene (that might be both dominant or both recessive). A **heterozygous** individual has two different forms of a gene, one recessive and one dominant. The combination of the alleles (two dominant, two recessive or one of each) is called the **genotype**.

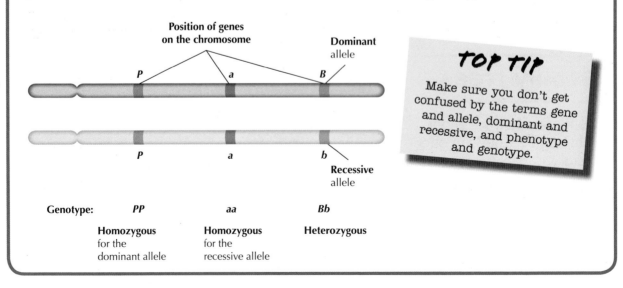

TOP TIP

Make sure you don't get confused by the terms gene and allele, dominant and recessive, and phenotype and genotype.

Usually, the upper case letter represents the dominant allele and the lower case letter represents the recessive allele.

One of the first people to suggest a mechanism for inheritance was Gregor Mendel. He worked with garden pea plants that varied in their flower colour, pea shape, height and other characteristics. He suggested that the gametes of each parent could only pass on one member of a pair of alleles. When the gametes fused, the phenotype of the new plants would depend on the various possible combinations of the alleles. The parents formed the **parental generation** and the next generation were called the **first filial** (or **F_1**) **generation** and the generation after that, the **second filial** (or **F_2**) **generation.**

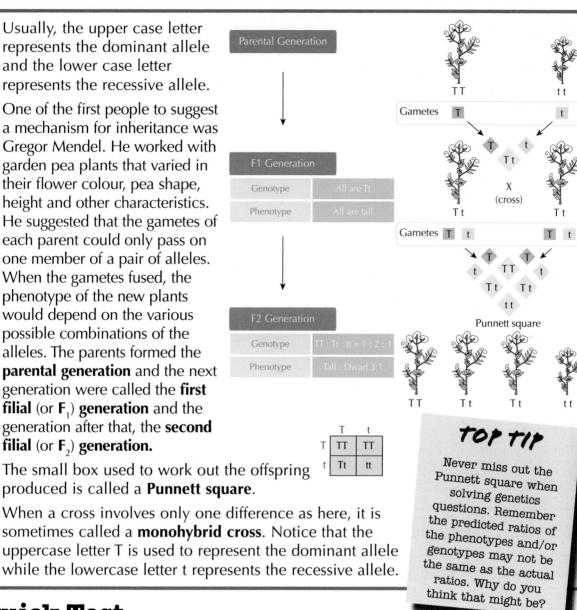

	T	t
T	TT	TT
t	Tt	tt

The small box used to work out the offspring produced is called a **Punnett square**.

When a cross involves only one difference as here, it is sometimes called a **monohybrid cross**. Notice that the uppercase letter T is used to represent the dominant allele while the lowercase letter t represents the recessive allele.

TOP TIP

Never miss out the Punnett square when solving genetics questions. Remember the predicted ratios of the phenotypes and/or genotypes may not be the same as the actual ratios. Why do you think that might be?

Quick Test

1. In pea plants, flower colour is determined by a gene that can exist in two different forms. The dominant (P) gives rise to a purple-coloured flower while the recessive (p) gives rise to a white-coloured flower. Two homozygous, differently coloured flowers were crossed to produce the F_1 generation. Two of these F_1 plants were crossed to produce the F_2 generation, which consisted of 270 purple-flowered plants and 90 white-coloured plants. Calculate the ratio (expressed as a simple whole number) of purple to white flowers.

2. Copy and complete this table by inserting the different human characteristics under the correct headings.

   height – weight – eye colour – hair length – hair colour

Type of variation	
Continuous	Discrete

3. Explain what is meant by 'polygenic inheritance'.

Transport systems – plants

Need for transport systems

Multicellular organisms have to exchange substances, such as oxygen and carbon dioxide, with their environment. However, because of their size, not all the body cells of a multicellular organism are in contact with their environment. While simpler, unicellular organisms can rely on diffusion, osmosis and active transport, as organisms get larger, their volume increases faster than their surface area, which just isn't able to match the demands of all the cells.

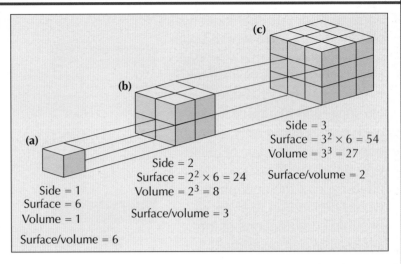

(a)
Side = 1
Surface = 6
Volume = 1
Surface/volume = 6

(b)
Side = 2
Surface = $2^2 \times 6 = 24$
Volume = $2^3 = 8$
Surface/volume = 3

(c)
Side = 3
Surface = $3^2 \times 6 = 54$
Volume = $3^3 = 27$
Surface/volume = 2

Consequently, transport systems in plants and animals have developed to allow exchange between the internal and external environments.

TOP TIP

Be aware that as an organism grows, its surface area to volume ratio is reduced.

Plant transport systems

In almost all multicellular organisms, water is needed for transporting materials.

Plant organs include roots, stems and leaves. The movement of water in plants from the roots to the leaves through the stem is called the **transpiration stream**, while the actual loss of water by evaporation from a plant is called **transpiration**. Most of the evaporative loss takes place through the leaves but some also takes place from the stems and flowers in plants.

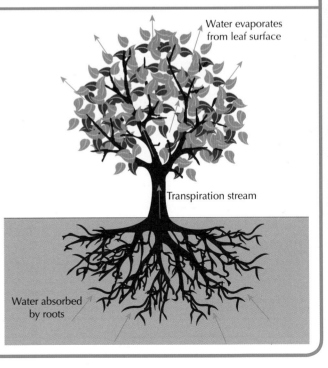

Water evaporates from leaf surface

Transpiration stream

Water absorbed by roots

TOP TIP

Remember water is needed for photosynthesis as well as transport in a plant.

Water loss from leaves takes place through tiny openings called **stomata** (singular **stoma**) found in the surface layers or **upper** and **lower epidermis** of the leaf. Covering the upper epidermis is a waxy layer called the **cuticle**, which helps cut down water loss. The opening and closing of the stomata is controlled by **guard cells** on either side. Below the upper epidermis is the **palisade mesophyll layer** consisting of tall, cylindrically-shaped cells, where most photosynthesis takes place.

> **TOP TIP**
>
> Palisade cells are packed with chloroplasts as an adaptation for their function in photosynthesis and their shape allows many to be packed close together.

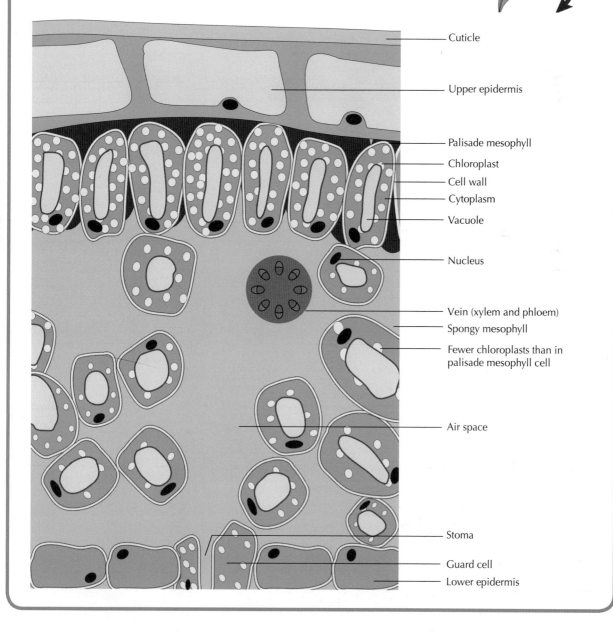

- Cuticle
- Upper epidermis
- Palisade mesophyll
- Chloroplast
- Cell wall
- Cytoplasm
- Vacuole
- Nucleus
- Vein (xylem and phloem)
- Spongy mesophyll
- Fewer chloroplasts than in palisade mesophyll cell
- Air space
- Stoma
- Guard cell
- Lower epidermis

The **spongy mesophyll layer** consists of irregularly shaped cells that do not fit together, allowing many air spaces between them. The air can circulate freely among these cells and so reach the other cells inside the leaf.

Water and minerals enter by osmosis through the roots via special **root hair cells**, which have a very large surface area for absorption.

Within the plant a transport system moves water up from the roots to the leaves in the form of narrow tubes called **xylem**, arranged near the outside of the stem. Close to the xylem is the **phloem** which transports food from the leaves to all parts of the plant.

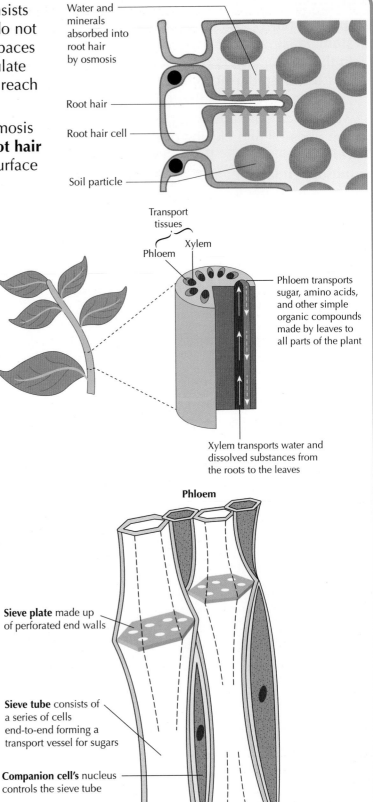

Water and minerals absorbed into root hair by osmosis

Root hair

Root hair cell

Soil particle

Transport tissues

Xylem

Phloem

Phloem transports sugar, amino acids, and other simple organic compounds made by leaves to all parts of the plant

Xylem transports water and dissolved substances from the roots to the leaves

Lignin

Xylem

To withstand changes in pressure as water travels through the dead xylem, these cells are strengthened with a material called **lignin**.

Phloem

Sieve plate made up of perforated end walls

Sieve tube consists of a series of cells end-to-end forming a transport vessel for sugars

Companion cell's nucleus controls the sieve tube

TOP TIP

Mature tall trees rely on lignin to support them.

As water evaporates from the surfaces of a plant, more water is drawn up to replace it. This creates a pull that is sufficient to support a very tall column of water, many metres in length.

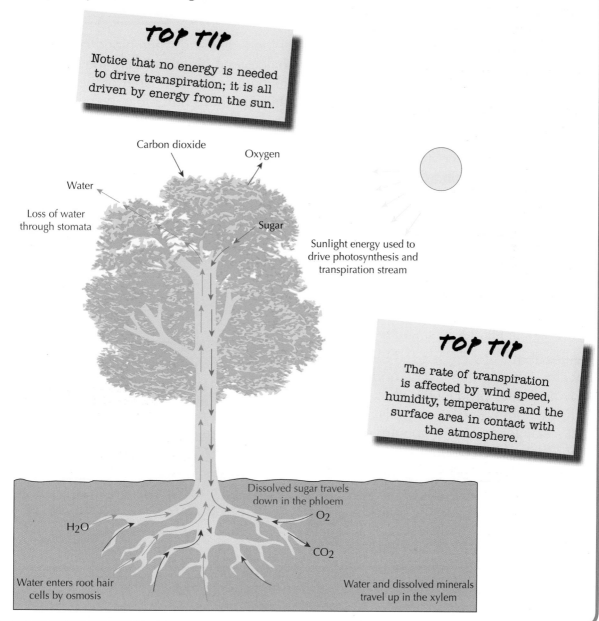

TOP TIP

Notice that no energy is needed to drive transpiration; it is all driven by energy from the sun.

Carbon dioxide

Oxygen

Water

Loss of water through stomata

Sugar

Sunlight energy used to drive photosynthesis and transpiration stream

TOP TIP

The rate of transpiration is affected by wind speed, humidity, temperature and the surface area in contact with the atmosphere.

Dissolved sugar travels down in the phloem

O_2

H_2O

CO_2

Water enters root hair cells by osmosis

Water and dissolved minerals travel up in the xylem

Quick Test

1. State three advantages of transpiration to a plant.

2. Explain why water will travel up the xylem more quickly on a dry day than on a humid day.

3. What is the function of lignin in xylem?

Transport systems – animals

Blood

In mammals the blood contains plasma, red blood cells and white blood cells. It transports nutrients, oxygen and carbon dioxide. **Red blood cells** are the most numerous of the cells found in the blood, making up about a quarter of the cells of the human body! A single drop of blood contains millions of red blood cells, which are constantly travelling through the body to supply cells with oxygen. They are unusual in having no nucleus or other structures such as mitochondria or ribosomes inside, leaving all the available space to contain as much of the protein **haemoglobin** as possible.

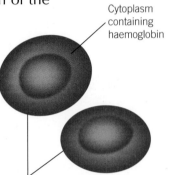

Cytoplasm containing haemoglobin

Haemoglobin is the carrier molecule that combines with oxygen to form **oxyhaemoglobin** in the lungs and then releases it at respiring cells. It also gives blood its red colour. Red blood cells are **biconcave**, which gives them a huge surface area to pick up and release oxygen across their membranes. The cells are also very flexible so they can squeeze through capillaries in single fire.

Biconcave discs with no nucleus, carry oxygen

After about 120 days, a red blood cell wears out, eventually dies and is replaced by new ones produced within material inside bones. Red blood cells are produced at a staggering rate of about two million per second in a healthy adult!

TOP TIP

The structure of a red blood cell is ideally suited to its function.

White blood cells are part of the body's defence or **immune system**. They destroy potentially harmful **pathogens** such as bacteria. There are two main types of white blood cell:

1. **Phagocytes** which can ingest and digest pathogens. This is called **phagocytosis** which involves engulfing and digesting pathogens.

2. **Lymphocytes** which can produce antibodies which specifically target pathogens. Each antibody is specific to a particular pathogen.

The liquid part of the blood in which cells are suspended is called **plasma**. The plasma transports glucose, amino acids, salts, enzymes, hormones and many other substances.

TOP TIP

Pathogens are disease-causing micro-organisms (bacteria, viruses, fungi).

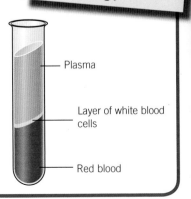

Plasma

Layer of white blood cells

Red blood

Pathway of blood

In mammals, blood transports vital nutrients and oxygen to every cell in the body and removes carbon dioxide from every cell. This is carried out by the **circulatory system**, which consists of blood moving in tubes called **vessels** and a pump called the **heart**.

The heart consists of four spaces called **chambers**.

The two upper chambers, or **atria** (singular **atrium**) receive blood from the body or the lungs while the two lower chambers, or **ventricles**, discharge blood to the lungs or to the body. Both the left and right sides of the heart have an atrium and a ventricle. **Valves** in the heart ensure that there is no backflow of blood and that blood flows in one direction only.

Blood flows into the heart through **veins** and out of the heart through **arteries**.

Like the heart, veins have valves to prevent backflow of blood. The main vein returning blood low in oxygen to the heart is the **vena cava** and the main artery taking blood high in oxygen from the heart is the **aorta**. The atria, which receive blood from veins, are thin-walled because they do not pump blood very far. However, the ventricles are both thick-walled because they have to pump blood under pressure to the lungs or to the rest of the body. The **pulmonary artery** carries blood low in oxygen from the right ventricle to the lungs while the **pulmonary vein** carries blood rich in oxygen from the lungs to the left atrium.

TOP TIP

The heart is really two pumps that beat as one unit. The left side deals with blood rich in oxygen while the right side deals with blood low in oxygen.

TOP TIP

Remember: <u>A</u>rteries carry blood <u>A</u>way from the heart and <u>V</u>eins have <u>V</u>alves.

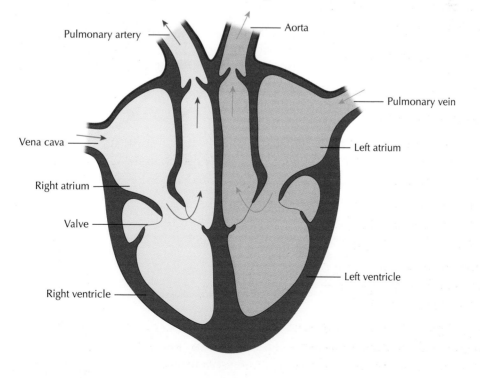

Pulmonary artery — Aorta — Pulmonary vein — Vena cava — Left atrium — Right atrium — Valve — Left ventricle — Right ventricle

The circuit of blood involves a return of deoxygenated blood to the right side of the heart from the body and then from the right side to the lungs. From the lungs, the oxygenated blood returns to the left side of the heart and then from the left side out to the body again.

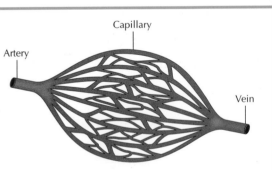

Blood vessels

All over the body **capillaries** are found, which connect arteries and veins. Capillaries are thin walled and have a large surface area, forming networks at tissues and organs to allow efficient exchange of materials. As blood flows through capillaries, an exchange of gases, nutrients and wastes takes place through their thin walls, which are only one-cell thick.

Capillary

Artery

Vein

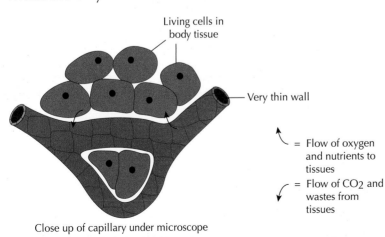

Living cells in body tissue

Very thin wall

= Flow of oxygen and nutrients to tissues

= Flow of CO_2 and wastes from tissues

Close up of capillary under microscope

TOP TIP

Capillaries have an enormous surface area for exchange of materials and are very close to every single cell in the body.

Arteries and veins differ in their structure and functions.

Artery	Vein
Close up of artery	Close up of vein
Thick muscular wall	Valve — Thin muscular wall
Transports blood away from heart	Transports blood to the heart
Carries blood rich in oxygen	Carries blood low in oxygen
Has a relatively narrow central canal	Has a wider central canal
Walls are very muscular	Walls have little muscle
Walls are very elastic	Walls are not very elastic
No valves are present	Valves are present
Blood is under high pressure	Blood is under low pressure
Ends in capillaries	Formed from capillaries
Pulse can be easily felt	Pulse not easily felt
Usually situated deep in body tissue	Usually situated nearer the skin

TOP TIP

The pulmonary artery and pulmonary vein are both exceptions to the general rule about arteries carrying blood rich in oxygen and veins carrying blood low in oxygen.

Quick Test

1. Explain how a red blood cell is adapted to its function.
2. State which chamber of the heart blood low in oxygen first enters.
3. Capillary walls are only one-cell thick. Explain how this aids exchange of materials.

Absorption of materials

Oxygen and nutrients

In order for aerobic respiration to take place, oxygen and nutrients from our food must be absorbed into the bloodstream. Through the bloodstream, blood carries these to all our cells. Waste materials, such as carbon dioxide, are removed from our cells and then also carried by the bloodstream.

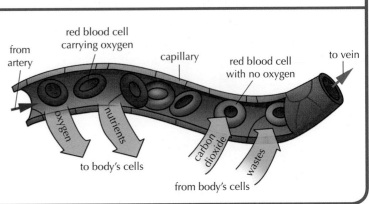

Capillary networks

Our body tissues contain capillary networks which allow these exchanges of materials, usually by diffusion, to take place at the cellular level. Such materials include oxygen, carbon dioxide, nutrients and waste.

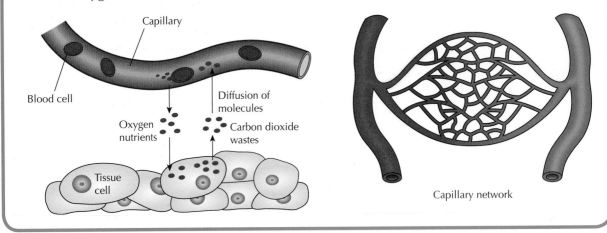

Capillary network

Features of surfaces involved in absorption

Surfaces which are involved in absorbing materials have a number of features in common:
- usually moist which allows gases to dissolve for easy diffusion
- very thin so that the distance travelled from one side to the other side of the surface is small
- large surface area to allow efficient and rapid exchange of materials
- covered by a dense network of capillaries to allow efficient exchange and transport of materials

got it? ☐ ☐ ☐

Absorption of materials

Lungs

The exchange of oxygen, needed for respiration, and carbon dioxide takes place in the lungs which are the gas exchange organs.

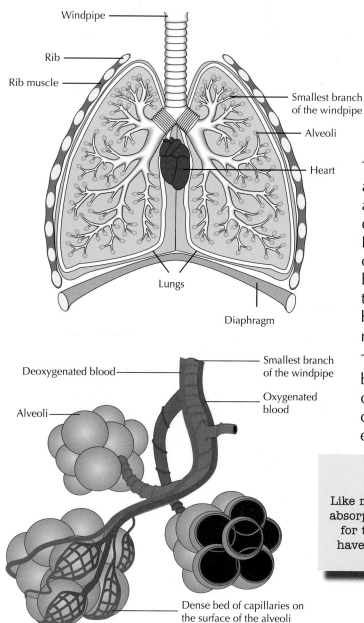

Windpipe

Rib

Rib muscle

Smallest branch of the windpipe

Alveoli

Heart

Lungs

Diaphragm

Deoxygenated blood

Smallest branch of the windpipe

Oxygenated blood

Alveoli

Dense bed of capillaries on the surface of the alveoli

TOP TIP

There are shared passages at the back of the mouth for food to enter the stomach via the gullet and air to enter the lungs via the windpipe.

The lungs have a huge surface area, which allows oxygen and carbon dioxide to diffuse quickly to and from the blood. However, as diffusion on its own is not rapid enough we have special muscles between the ribs, which cause air to be pulled in and pushed out much more quickly.

The ends of the smallest branches of the windpipe consist of tiny little air sacs called **alveoli** where gas exchange takes place.

TOP TIP

Like most surfaces involved in exchange or absorption of materials, alveoli are adapted for their function. They are thin-walled, have a dense blood supply, a moist lining and a huge surface area.

Blood flows through the lungs in capillaries, which are in very close contact with the alveoli. Oxygen diffuses from the alveoli into the blood in the capillaries and carbon dioxide diffuses from the blood in the capillaries into the alveoli. Since there are millions of alveoli in each lung, the overall surface area for gas exchange at the cellular level is enormous.

Air is moved in and out of the breathing system during inhalation and exhalation using the rib muscles and the **diaphragm**.

Absorption of nutrients from food

After we have eaten, food is broken up into small pieces until, eventually, the large molecules of fat, protein and carbohydrate are fully broken down into small molecular products to be transported in the blood and used by every cell in the body. Therefore food is both mechanically and chemically broken down from large pieces into small pieces, and large molecules into small molecules.

Mechanical and chemical digestion starts in the mouth where our teeth break up the food into small pieces. Saliva is added, which helps to make the food slide down the **oesophagus** into the stomach.

Saliva contains an enzyme called **amylase** which starts the breakdown of starch. It also contains mucus to make food slide down the oesophagus and easier to swallow.

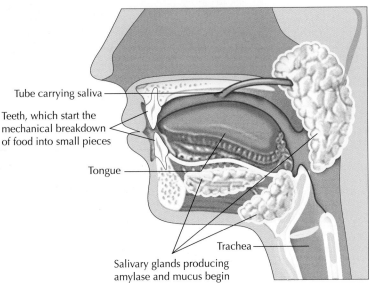

Tube carrying saliva

Teeth, which start the mechanical breakdown of food into small pieces

Tongue

Trachea

Salivary glands producing amylase and mucus begin chemical digestion

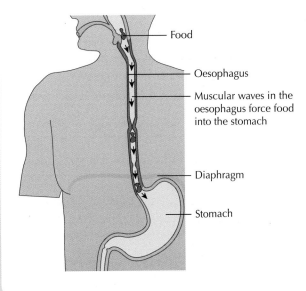

Food

Oesophagus

Muscular waves in the oesophagus force food into the stomach

Diaphragm

Stomach

As food travels along the digestive system, different parts contribute towards converting the food into a form that can be used by body cells.

After chewing food and mixing it with saliva, food is passed down the oesophagus into the stomach by muscular waves.

TOP TIP

Chewing food exposes a large surface area for enzymes to act on.

TOP TIP

Swallowing is an example of a reflex action.

In the stomach, digestive enzymes and hydrochloric acid are added to food to continue the breakdown process. The partially digested food now enters the **small intestine** to complete digestion and allow **absorption** to take place.

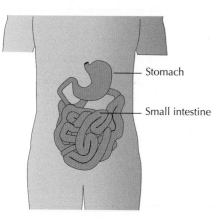

Stomach

Small intestine

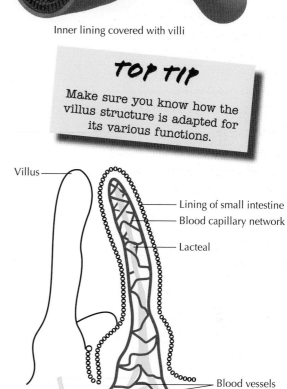

Inner lining covered with villi

TOP TIP

Make sure you know how the villus structure is adapted for its various functions.

The small intestine is highly adapted for the process of absorption. Its inner lining is covered with many tiny finger-like structures called **villi**, which hugely increase the surface area. Villi have a number of structural adaptions making them well suited to their function:

- Covering is only one-cell thick so that digested food has a short distance to travel to enter the blood
- Dense blood supply in the form of many capillaries for absorption of sugars, amino acids, water, minerals and many vitamins.

Villus

Lining of small intestine

Blood capillary network

Lacteal

Blood vessels

In the centre of each villus is found a **lacteal** which absorbs the end products of the digestion of fats called **fatty acids** and **glycerol**.

Blood carries all the products of digestion, except those of fat, towards the liver. Digested fats are carried in lymphatic vessels in lymph towards the heart where they enter the bloodstream.

Quick Test

1. Name two vital substances, transported in the blood, which are needed by the heart muscle.

2. State three features common to surfaces which absorb materials.

3. State two substances absorbed into the lacteal of a villus.

Revision Questions

Section A

1. Which of the following descriptions is correct?

	Tissue	Organ	System
A	Xylem	Phloem	Photosynthetic
B	Stem	Leaf	Transport
C	Sperm	Testis	Reproductive
D	Nerve	Brain	Nervous

2. Which of the following pairs of responses are both examples of reflex actions?

   A Sneezing and standing up

   B Coughing and scratching itchy skin

   C Breathing and speaking

   D Vomiting and blinking

3. The diagram below shows a human brain.

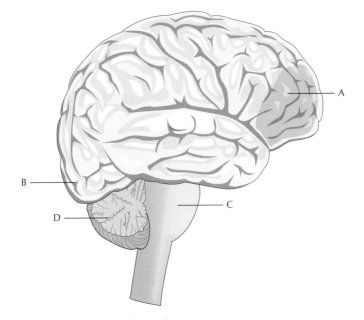

   State which part regulates the heartbeat.

4. Which one of the following properties of glycogen is not correct? Glycogen

   A can be used directly by the brain as a quick source of energy

   B being insoluble, does not set up any osmotic effects

   C is an ideal form of energy storage

   D is a large molecule formed from many smaller units.

5. Which one of the following pieces of apparatus could be used to compare the relative amount of carbon dioxide in inhaled air with that in exhaled air by breathing in and out at the mouthpiece shown as M?

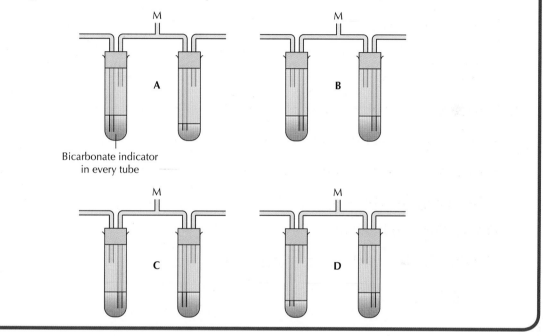

Bicarbonate indicator
in every tube

Section B

1. Copy the information below and connect the terms on the left to the correct statements on the right by means of arrow-headed lines.

Diploid	Number of chromosomes in sperms
Haploid	Fertilised egg
Testis	Generalised name for a sex cell
Gamete	Number of chromosomes in plant body
Zygote	Structure where sex cells are manufactured

[2]

2. (a) Explain what is meant by single-gene inheritance. [1]

   (b) State two examples of single-gene inheritance. [1]

3. In a particular type of dog, the coat may be a plain pattern or a spotted pattern. Plain pattern (S) is dominant while spotted pattern (s) is recessive.

   (a) A spotted dog is crossed with a heterozygous plain-patterned dog.
      (i) State the possible phenotypes of the offspring. [1]
      (ii) State the possible genotypes of the offspring. [1]

   (b) A heterozygous plain-patterned dog is mated with another heterozygous plain-patterned dog.

      Calculate the genotypic ratio of the offspring. [1]

4. Copy and complete the following paragraph by inserting the correct words into the spaces provided.

   Water is moved from the roots to the leaves in the _____ and dissolved sugar is moved from the leaves to the roots in the _____. [1]

5. Copy and complete the following table. [2]

	Artery	**Vein**	**Capillary**
Relative diameter	Small		
Wall		Thin muscular	
Valves			Absent

6. A student investigated how feedback might improve a person's ability to draw a line of a certain length. Three volunteers were separately first shown a line that was exactly 8 cm long, drawn on paper. Next, each volunteer in turn was seated as shown in the diagram to the right and asked to draw a line of 8 cm on five consecutive sheets of paper that were behind a wooden screen. The volunteers were asked not to touch the table as they drew the lines. Before the sixth attempt, the volunteers were told the average of their previous five attempts and then given a further five attempts. The data obtained are shown in the table.

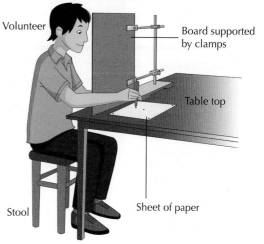

Attempt	Length of lines drawn (cm)		
	Volunteer 1	**Volunteer 2**	**Volunteer 3**
1	8·1	6·0	4·5
2	7·9	6·7	5·5
3	7·7	8·3	7·5
4	8·2	8·2	8·5
5	7·6	8·3	5·0
Average			
6	7·9	6·2	5·5
7	7·9	6·6	5·8
8	8·1	8·3	9·0
9	8·1	8·1	7·8
10	8·0	7·8	7·9
Average			

(a) Copy and complete the table by calculating the average results for each volunteer. Then create a bar chart showing the average results of the first and second attempts of the three volunteers. [4]

(b) Name the dependent variable in this investigation. [1]

(c) The student concluded that feedback improved a person's ability to draw a line of a certain length.

Suggest two reasons why this was not a valid conclusion. [2]

(d) Explain how the reliability of the results could be improved. [1]

(e) State two ways in which the student could improve the design of this investigation. [1]

Ecosystems

Ecosystems

In an environment, such as a woodland or pond, individual living things form **populations**: a group consisting of only one **species**. Together, populations form different **communities** and with their non-living environment communities form an **ecosystem**. Ecosystems collectively form the **biosphere**, the part of the Earth's surface and atmosphere within which life can exist.

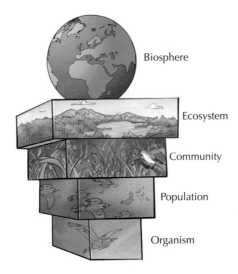

Many different types of ecosystems are possible such as:

- desert
- forest
- pond
- river
- estuary
- heather moorland
- arctic tundra
- coral reef.

> ### TOP TIP
> Make sure you know the distinction between the terms habitat, population, community and ecosystem.

Within ecosystems, green plants can make their own food by the process of photosynthesis, converting sunlight energy into chemical energy. They are known as **producers** because they are the source of food for animals. Since animals cannot make their own food, they are known as **consumers**. Simple feeding relationships can be shown in the form of a **food chain** which can have several links.

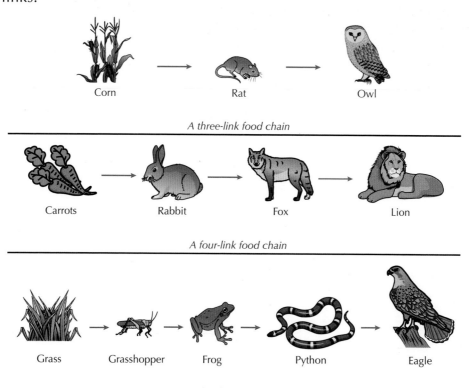

A three-link food chain

A four-link food chain

A five-link food chain

The arrows in a food chain indicate the flow of energy, always pointing to the organism which is feeding on the previous one. Animals which eat only plants are called **herbivores.** Animals which eat only other animals are called **carnivores**. Some animals eat both plants and animals and are called **omnivores**.

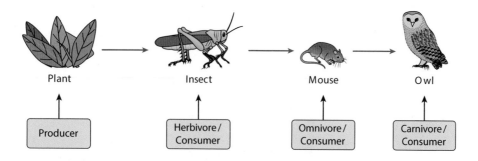

In the above diagram, the insect is the **prey** of the mouse which in turn is the prey of the owl. The mouse is the **predator** of the insect and the owl is the predator of the mouse.

In nature, food chains usually interlink with each other to form **food webs**.

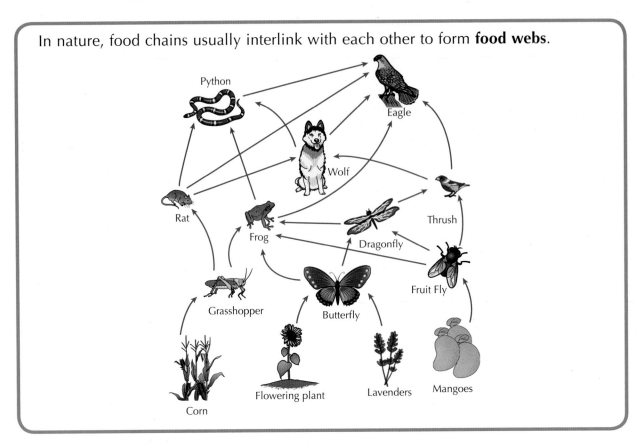

Niches

In their environment, living things perform particular roles called **niches**. An organism's niche relates to the resources it requires in its ecosystem, such as light and nutrient availability and its interaction with other organisms in the community. A niche involves competition and predation and the conditions it can tolerate such as temperature.

Branches and leaves:
bees, wasps, moths, squirrels, blue tits and hawks

Trunk:
insects and larvae

Root and litter zone:
bacteria, earthworms, woodlice and fungi

Competition

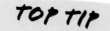

Resources such as space, food, water, light, mates, and shelter are not without limit on Earth; animals and plants have to compete for these, especially if any are in short supply. If the competition is between members of the same species this is called **intraspecific competition.** If the competition is between members of different species this is called, **interspecific competition**. Competition helps prevent populations of animals and plants becoming too large to be supported by their environment.

Grasshoppers often feed in large swarms that are made up of the same species and all the individuals compete for the same food source.

Animals, such as robins, may compete for the same space known as their **territory** and make this known by their behaviour patterns. For example, a robin's red-coloured breast establishes territorial rights by making them more visible and attractive to potential mates.

In many parts of Scotland, the grey squirrel has reduced the numbers of the native red squirrel. These two different species compete for the same resources but the grey squirrel is more successful.

As cheetahs and lions, which are different animal species, both feed on similar prey, such as gazelles, the presence of one has a negative impact on the other because they both compete for the same food resource.

In any large, well-established forest, many different species of plants will compete for shared resources such as:

- light
- carbon dioxide
- water
- nutrients.

Plants in such a well-established forest have different heights as they receive varying amounts of these shared resources. The different species will compete vigorously; those that are better suited to obtain them will grow best and outperform the weaker plants, which may grow poorly or in the long term be completely wiped out.

Sometimes a non-native plant species can invade and start to dominate a forest, such as in Hawaii when ground-dwelling ferns became established as the dominant plant form.

TOP TIP

Interspecific competition is less intense than intraspecific competition because different species do not need exactly the same resources.

Quick Test

1. Explain the difference between interspecific and intraspecific competition.
2. State the term used to describe where a plant or animal lives.
3. Explain what a 'community' is.

Distribution of organisms

Biotic and abiotic factors

Plants, animals and micro-organisms are found everywhere on planet Earth. A living thing lives in a **habitat.** The range of living things that are found in an environment is called **biodiversity** and is often delicately balanced. It can be affected by different factors that may be **biotic** or **abiotic**.

Biotic factors include:

- competition for food, space, mates
- the number of predators
- disease
- grazing.

Abiotic factors include:

- temperature
- humidity
- pH
- light intensity
- rainfall
- wind.

Humans can influence their local, national and global environments by such activities as:

- pollution of the air or water
- destruction of natural habitats
- exploitation of animals and plants.

> ***TOP TIP***
> Think of the different ways in which humans can destroy natural habitats.

Measuring abiotic factors

> ***TOP TIP***
> Remember always to use the phrase 'light intensity' and not just 'light' when describing this abiotic factor.

Measuring abiotic factors is a vital part of building up a picture of how an ecosystem is working.

Light meters can be used to measure light intensity.

They are usually placed on the ground and a reading is taken from a scale. It is important to give the meter a little time to stabilise and also ensure no accidental shading takes place. The readings should be taken, as far as possible, at the same time to avoid the effects of passing clouds over one site.

Combined meters measure soil water content and pH. They have an electronic probe that is pushed into the soil. It is important to push the probe in to the same depth each time (usually there is a special mark to aid this) and it must be cleaned between readings. The meters need time to stabilise before readings are started.

A number of different types of thermometers, including some digital devices, measure air, soil and water temperatures accurately. The meters require time to stabilise between readings and, if a probe is attached, it must be pushed in to the same depth each time and cleaned between readings.

Meters now can easily be attached to software to capture data over a period of time and link up with a computer. As with all measurements, repeated sampling is required to ensure reliable results.

TOP TIP

Think why the probe needs to be cleaned between readings.

Sampling techniques

If a groundsman wanted to know how many dandelions were growing in a football pitch, it would be impractical to try to count every single dandelion. Instead, he might select a number of small areas, a process called **sampling**. All the samples would be of the same size and he would count the number of dandelions in each. From this he could calculate an average number of dandelions in one of these small areas and then multiply this to the actual area of the football pitch.

There are a number of techniques which can be used on their own or together when sampling.

A **quadrat** is a commonly used simple piece of apparatus, usually made of metal, with a known area, often 0.25 m^2. This is laid randomly on the ground and the plants or animals of interest are counted. From this data, an average number of the organisms is obtained. Knowing the actual area being investigated, it is then possible to estimate the number of these organisms present.

Quadrats can be used on land but also along shores and under water.

TOP TIP

Quadrats can also be made of wood or string.

Surface-dwelling animals can be sampled using a **pitfall trap**. This is usually a jar or tin can sunk into the soil overnight. The animals fall into the container and can be studied the following day.

In studying an ecosystem, it is necessary to have data related to the numbers of organisms present but it is rarely possible to count all of these. It might be easy to do this with large animals such as giraffes in a restricted area but this would not be possible to do with, for example, ants in the same restricted area. The numbers of trees in a forest are relatively easy to count because they don't move!

Samples must be:

- completely random to avoid any bias
- truly representative of the whole area being studied.

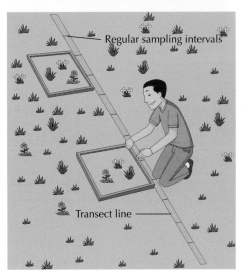

Regular sampling intervals

Transect line

To enhance the reliability of the results, many samples must be taken, not just a few.

Using better apparatus will improve the accuracy but not the reliability of the results.

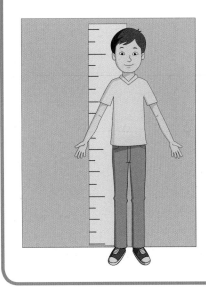

Suppose a student wanted to estimate the average height of male students in first year at her school using a handheld measuring tape. If she measures only five students out of the total population of 150, this will not be very reliable as she may have five students who are all unusually tall. By increasing the sample size, her results become increasingly more reliable. If she changed the tape for a digital height measurer, this would make each individual result more accurate but not necessarily make her overall results any more reliable.

Evaluating limitations and sources of error in sampling techniques

Any one technique used in ecology will have its own limitations and sources of error, which is why ecologists use a number of different strategies in their studies. Consider the two techniques mentioned earlier, the pitfall trap and the quadrat. Here are some of the limitations and sources of error, as well as suggestions as to how the effect of these can be reduced.

> **TOP TIP**
>
> Think of the other techniques mentioned in a similar way.

Sampling technique	Limitation and error source	Suggested ways of reducing effect
Pitfall trap	Sample obtained may not be representative of the whole area	Take as many samples as possible from as many traps as possible
	Trapped animals may eat each other	Do not leave the traps for long periods before emptying
	Animals may avoid the trap	Ensure it is well blended in with the natural habitat
	Other, larger animals may dip into the trap and eat the small animals	Make sure the trap is suitably covered with a protective lid or stone
Quadrat	Sample obtained may not be representative of the whole area	Take as many samples as possible from as many quadrats as possible
	Some organisms may fall partially in or out of the quadrat	Ensure the method of counting organisms in or out is established and consistently applied to each square
	Samples may not be random	Use some way of ensuring the quadrats are not 'placed' in any patterned way

Paired-statement keys

To identify animals and plants studied in ecosystems, biologists use devices called **keys**. The most common type of key uses statements arranged in pairs. The paired statements usually give some feature that may or may not be present leading on to more statements and ultimately naming the plant or animal.

1. Wings present ....Bee
   Wings absent ....Go to 2
2. Eight legs...House spider
   Six legs ....Go to 3
3. Narrow waist ....Ant
   Broad waist..Flea

Effect of biotic and abiotic factors on biodiversity

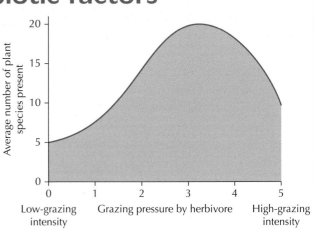

Herbivores, such as zebra, antelope, etc., feed by grazing on plant life. The effect this grazing has on biodiversity depends on the level of the grazing. At low-grazing intensity, dominant plants may prevail making it difficult or impossible for more delicate forms to survive, thus reducing biodiversity. At high-grazing intensity, perhaps due to increased population numbers, less and less of the plant life can survive as the herbivores compete for shared resources. This also decreases biodiversity. With a balanced interaction between grazers and their food sources, native plants and grasses will prosper because they don't have to compete with the more aggressive surrounding non-native species. Grazers have a positive effect on biodiversity, too. They often push seeds into soft soil with their feet and these seeds germinate later. They also fertilise the soil with their waste products.

Recent work has shown that predation is a critical factor in maintaining high levels of biodiversity. For example, the decline in certain large cats in some national parks has allowed deer and similar mammals to reproduce in large numbers, with the loss of many plant species that are, in turn, eaten by these animals. The loss of plant species brings about soil erosion, while pollinators, such as bees and butterflies, may no longer have suitable sources of

pollen. Similarly, populations of wild foxes and cats may increase dramatically if their natural predators reduce in numbers; this in turn may drastically reduce the numbers and types of their food.

Fish are generally very sensitive to changes in the pH and temperature of the water. Most aquatic environments have a pH that is nearly neutral, around 7. If this is lowered or increased by pollution it can interfere with the reproductive lifecycles of fish as well as promoting the invasive growth of plants which cannot grow at a pH of 7, reducing the number of fish species. Some processes, such as generating power, use vast quantities of

water as a coolant. This warmer water, when discharged back into rivers and streams, can elevate the temperature of rivers and streams where fish thrive. A similar effect is caused by global warming. In water that is warmer than normal, fish mature quicker but produce fewer offspring. The amount of dissolved oxygen in warm water is less and this, coupled with an increased rate of respiration, causes many fish eventually to die out.

Indicator species

If an ecosystem changes, the organisms that normally live there will be affected. Some organisms are particularly sensitive to even small changes in variables such as pH, temperature, light intensity, etc. They may either decrease or increase in numbers.

Lichen is an association between two organisms, an alga and a fungus. It is very sensitive to levels of sulphur dioxide pollution and, in general, will not tolerate even a small rise in such pollution. The numbers and appearance will change considerably if, for example, levels of the sulphur dioxide gas rise. Rock surfaces and tree barks with a dense and healthy lichen growth indicate good air quality.

Otters will not thrive in water that is polluted. They are carnivores and so, if there is pollution in their habitat, they will bioaccumulate any pollutants present in their prey.

TOP TIP

When bioaccumulation occurs, animals at the top of a food chain experience the more severe effects of the pollution.

Certain maidenhair ferns will only grow in habitats that have a mineral called serpentine. Thus the growth of these ferns indicates the presence of this mineral and a particular type of soil substrate.

Some woodpeckers are found only in well-established and stable woodlands. Large numbers of these birds indicate a good quality habitat. If the health of this animal starts to decline, it is an indication that the ecological health of the habitat is in poor condition for some reason.

Water containing a large number of the early (nymph) stages of stoneflies and mayflies indicates high levels of oxygen in the water. However, if these are found to be absent and bloodworms are present in high numbers, this is a good indication of low levels of oxygen and high levels of pollution, often by sewage.

Organisms such as these, that can indicate the condition of a particular habitat or environment, are called **indicator species**. They are a very powerful monitoring tool for indicating the levels of certain variables and how these levels change over time because they are very sensitive to small increases or decreases in these variables.

TOP TIP

An indicator species by its presence or absence indicates the quality and/or levels of pollution in the environment.

TOP TIP

Remember, the indicator species may be absent or present as an index of ecological health.

Quick Test

1. A student decided to study the invertebrates living on the branches and leaves of a chestnut tree and an elder tree. He used the method shown.

   (a) Suggest if his results would be representative of all the different kinds of invertebrates living on this tree. Explain your answer.

   (b) State two variables he would need to ensure he kept the same throughout to make his comparison between the trees valid.

2. Name a technique that might be used to measure an abiotic factor. For this technique, identify one possible source of error and state how this error might be minimised.

3. Put the following factors under the correct column heading in the table:
   temperature – light intensity – predation – humidity – wind – salinity – disease – soil moisture – food availability – space

Abiotic factor	Biotic factor

Photosynthesis

Introduction

One of the most fundamental differences between animals and plants is how they obtain their food. Whereas animals need to have their food 'ready-made' in the form of other animals or plants, plants use the energy of the sun to produce their own food in the form of the **carbohydrate**, **glucose** (a type of sugar). This process is called photosynthesis and takes place in the chloroplasts of green plant cells. The green chlorophyll in the chloroplasts traps the light energy. Oxygen is released and diffuses through the stomata to the atmosphere. Photosynthesis can be summarised by the following word equation:

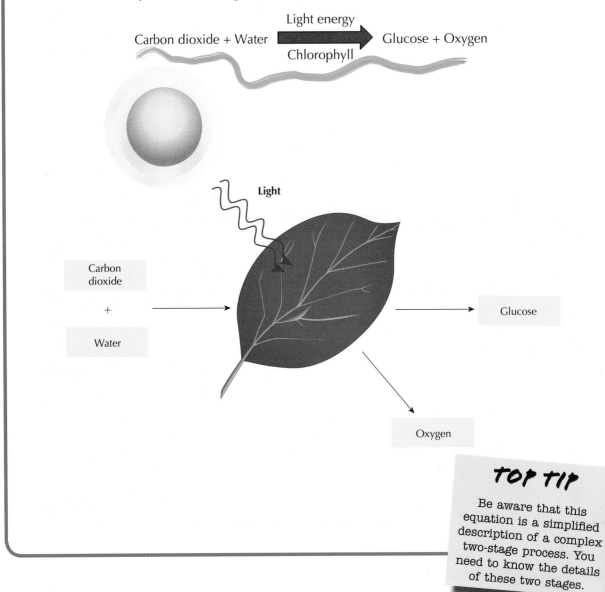

Carbon dioxide + Water →(Light energy / Chlorophyll)→ Glucose + Oxygen

Light

Carbon dioxide
+
Water

Glucose

Oxygen

TOP TIP

Be aware that this equation is a simplified description of a complex two-stage process. You need to know the details of these two stages.

Chemistry of photosynthesis

Photosynthesis has two stages, both consisting of many enzyme-controlled reactions. The first stage needs light energy, which is trapped by chlorophyll and used to produce **ATP**. Water, taken in by the roots of the plant, is split into hydrogen and oxygen using the energy of sunlight. The hydrogen is used in the second stage, which does not need light, while the oxygen diffuses out of the cell. In the second stage, the hydrogen combines with carbon dioxide to form glucose and the energy to drive this comes from ATP. This is called **carbon fixation**.

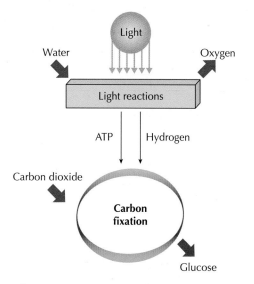

The glucose produced by photosynthesis can be used in many different ways by the plant. For example, the chemical energy present in glucose is available for respiration or the glucose can be converted into plant products such as **starch** or cellulose. When light is not available, the plant still needs energy and it obtains this by breaking down the glucose. Glucose can also be converted into **fats** and **oils** as well as **proteins**.

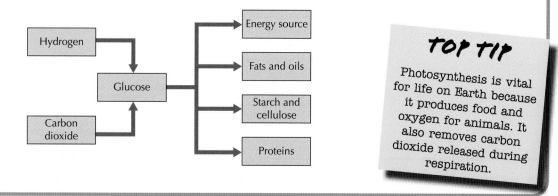

TOP TIP

Photosynthesis is vital for life on Earth because it produces food and oxygen for animals. It also removes carbon dioxide released during respiration.

Limiting factors

Since cell growth is closely linked to how quickly photosynthesis takes place, anything that affects this process will in turn speed up or slow down the growth of the plant. One way of measuring the rate of photosynthesis is to measure the volume of oxygen produced in a fixed time by an aquatic plant. It is also possible to use the increase in the dry mass of a plant or volume of carbon dioxide taken in within a fixed time.

Three important factors will affect the rate of photosynthesis:

1. Temperature

2. Light intensity

3. Carbon dioxide concentration.

At any one time, only one factor limits the rate of photosynthesis; this is therefore called the **limiting factor**.

As the light intensity increases from point A to B, the rate of photosynthesis increases until it starts to level off and then stays constant at point C. At this value of light intensity some other limiting factor, such as temperature or carbon dioxide concentration, is preventing an increase in the rate of photosynthesis.

As the concentration of carbon dioxide increases, the rate of photosynthesis increases but eventually levels off as some other limiting factor, such as light intensity or temperature, prevents an increase in the rate of photosynthesis.

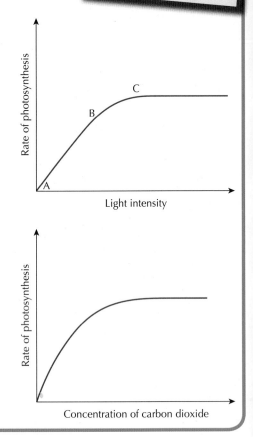

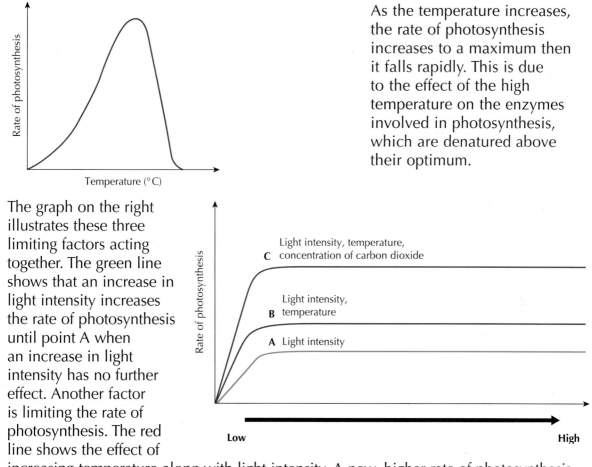

As the temperature increases, the rate of photosynthesis increases to a maximum then it falls rapidly. This is due to the effect of the high temperature on the enzymes involved in photosynthesis, which are denatured above their optimum.

The graph on the right illustrates these three limiting factors acting together. The green line shows that an increase in light intensity increases the rate of photosynthesis until point A when an increase in light intensity has no further effect. Another factor is limiting the rate of photosynthesis. The red line shows the effect of increasing temperature along with light intensity. A new, higher rate of photosynthesis is obtained but this also reaches a maximum at point B, when no further increase in light intensity or temperature increases the rate of photosynthesis. The blue line shows the effect of increasing the concentration of carbon dioxide as well as light intensity and temperature. Again, a new higher rate of photosynthesis is obtained that reaches a maximum at point C when no further increase in any of these three limiting factors will increase the rate of photosynthesis.

Removing limiting factors increases the rate of photosynthesis. This, in turn, produces more glucose to supply more energy for plant growth.

TOP TIP

This type of presentation with more than one limiting factor is often asked in examinations.

Quick Test

1. Explain what is meant by 'carbon fixation'.
2. State two uses of glucose made in photosynthesis.
3. In studying the effect of light intensity on the rate of photosynthesis, which of the following limiting factors must be controlled?
   a) temperature and carbon dioxide concentration
   b) oxygen concentration and temperature
   c) temperature and light intensity
   d) oxygen concentration and water availability.

Energy in ecosystems

Energy loss

To survive, all living things need energy. While animals obtain their energy from respiration, plants rely on the sun. Only a very tiny fraction of the total energy of the sun is actually absorbed by plants and used in photosynthesis.

As plants are eaten by animals and they, in turn, are eaten by other animals, about 90% of the energy gets lost as heat, movement or is trapped in undigested materials. When an animal or plant dies, the remains are broken down by bacteria, fungi and invertebrates called **decomposers** and a lot of energy is lost as heat to the atmosphere. Only a very small quantity is used for growth and is therefore available at the next level in a food chain.

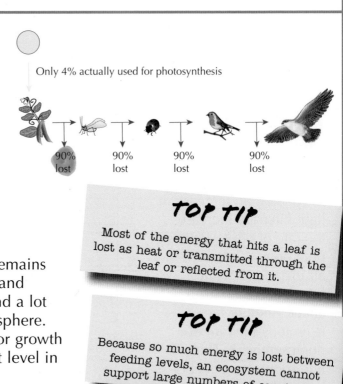

Only 4% actually used for photosynthesis

90% lost   90% lost   90% lost   90% lost

TOP TIP

Most of the energy that hits a leaf is lost as heat or transmitted through the leaf or reflected from it.

TOP TIP

Because so much energy is lost between feeding levels, an ecosystem cannot support large numbers of carnivores.

Pyramids of numbers and energy

Pyramid diagrams can be used to represent changes in the variables associated with ecosystems. Two different variables can be represented in this way:

1. Numbers
2. Energy.

As energy is lost at each feeding level, the number of organisms normally decreases proportionally. This change can be represented by a **pyramid of numbers**.

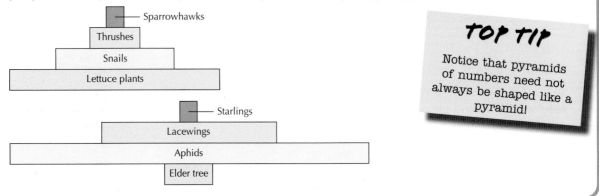

Sparrowhawks
Thrushes
Snails
Lettuce plants

Starlings
Lacewings
Aphids
Elder tree

TOP TIP

Notice that pyramids of numbers need not always be shaped like a pyramid!

A more valid way of representing the changes that take place from one feeding level to the next is the **pyramid of energy**.

Such pyramids are more difficult to construct; they require a lot of data collected over long periods but give a better picture of how an ecosystem is working.

10 J
snake

100 J
mouse

1,000 J
caterpillar

10,000 J
sunflower

1,000,000 J of sunlight

1 J
owls

10 J
shrews

100 J
insects

10,000 J
grass

1,000,000 J of sunlight

TOP TIP

Representing the energy changes always creates a true pyramid because energy is always lost moving from one feeding level to the next.

Quick Test

1. The diagram opposite shows a pyramid of numbers. State the level which represents the producers.

2. The organisms at the top of a pyramid of numbers are usually:

   a) least numerous and smallest

   b) most numerous and smallest

   c) least numerous and largest

   d) most numerous and largest

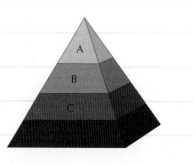

3. An ecosystem receives 6 000 000 units of energy from the sun. Of this energy, only 5% is used in photosynthesis. Calculate the energy which is lost to the producers.

Food production

Increasing human population

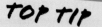

TOP TIP

Make the link with Darwin's observation that organisms produce more offspring than can possibly survive in the section on Evolution of species.

A stable environment can only support a certain number of a particular organism. As adults die they are replaced by offspring, not all of which survive.

A number of factors keep the population in check such as:

- Food and water availability
- Numbers of predators
- Disease
- Competition.

The concept that a stable environment can only sustain a limited number of a particular organism also applies to humans. Present estimates of the human population put the number at around 7·5 billion people with this potentially rising to over 9 billion by 2050. We are not subject to the same regulatory factors as other species because we can control their impact. For example, humans no longer are subject to animal predation!

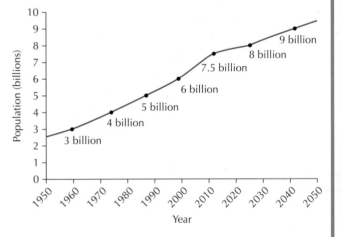

The huge increase in the human population continues to require more efficient ways of producing food to meet the demand. This increased demand can involve the use of fertilisers and pesticides. Fertilisers provide chemicals such as nitrates which increase crop yield. Plants and animals which reduce crop yield can be killed by pesticides.

In 2009 the United Nations claimed it would be necessary to increase food production by over 70% over the next 40 years to feed the world's increasing population. Humans have been able to meet the food demands of an increasing population by a number of combined strategies such as:

- Genetic diversity: growing a variety of different strains of crops that have similar features but originate from different parents. Such plants can be bred to increase yields and have natural resistance to pests and diseases. Genetic diversity is important in cattle and fish breeding programmes. Increased yields in animal food production and their products, such as milk, have made huge strides forward over the last 30 years or so.

- Improvement in soil activity: the addition of organic material encourages a wider variety of different organisms, increasing biodiversity. A wider range of plants grow on such nutrient-rich soils. Rotating crops prevents the build up of specific pests. Keeping soils well aerated prevents anaerobic conditions developing and improves water drainage.

- Employing modern farming techniques: the use of knowledge-based strategies for obtaining a more realistic potential for land used to grow crops; using the best balance of feed for animals to obtain the highest yield.

- The use of genetically modified organisms: animals and plants can be genetically modified to increase their yield, making them easier to grow and harvest, be more resistant to disease, and having increased nutritional profiles.

Nitrates

Nitrates are essential for amino acid production. Amino acids are synthesised into plant proteins which are, in turn, consumed as food by animals. This food can be directly in the form of plants or other animals which eat plants. Nitrates dissolve rapidly in soil water and are absorbed in plants. If a soil is low in nitrates, growers can add nitrate in the form of **fertilisers**, either natural, such as animal manure,

or artificial, in the form of chemicals. This will increase the yield obtained by encouraging plant growth. However, fertilisers should not be applied in excess so that they can be washed out of the soil and into waterways.

Algal blooms

Bodies of water such as lakes, reservoirs, rivers and streams will contain plant life which includes **algae**. These are microscopic, photosynthetic organisms that lack the usual structures associated with plants such as stems, roots and leaves. Like any plant, its growth is affected by levels of sunlight, carbon dioxide availability, suitable temperature, and nutrients such as phosphorus and nitrogen. Nitrates supply nitrogen to plants to produce amino acids which are synthesised into plant proteins. Animals eat plants or other animals to obtain their amino acids to make animal proteins.

TOP TIP

Sometimes algal blooms can occur naturally due, for example, to changing weather patterns bringing up nutrients from deep waters.

These nutrients are found as phosphates and nitrates, which, on occasion, can be **leached** by rainwater from soil that has had fertilisers added. If these nutrients find their way into bodies of water, they will cause the algae to multiply rapidly to form an **algal bloom**.

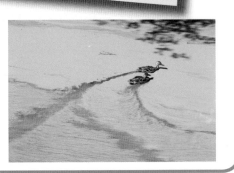

Associated with modern-day farming methods and living styles, there has been an increase in this nutrient loading of water. Some causes include:

- Leaching of inorganic fertiliser (containing nitrates and phosphates)
- Leaching of organic fertiliser manure (containing nitrates and phosphates) from intensive farming
- Leaching from erosion (following mining, construction work or poor land use)
- Discharge of detergents (containing phosphates)
- Discharge of partially treated or untreated sewage (containing nitrates and phosphates).

> **TOP TIP**
>
> Many supermarkets have withdrawn washing products with high levels of phosphates.

The enhanced growth of the algae (and other aquatic plants) reduces the levels of dissolved oxygen in the water when dead plant material decomposes and becomes food for bacteria which increase greatly in number. The bacteria use up large quantities of oxygen. The lowered levels of oxygen in turn then affect organisms, such as fish and other aquatic life forms, which may die. Algal blooms can also reduce light levels inhibiting photosynthetic activity.

> **TOP TIP**
>
> On land, excess nutrients can leave the soil and don't cause too much damage, but in bodies of water, such excess nutrients have nowhere to go.

Some algal blooms can pose a threat to land animals including humans. For example, if humans eat fish or shellfish from water polluted by certain toxic algae, they may become ill. Even dogs licking the water from a reservoir with a heavy algal bloom have been known to die. Treated water taken from such reservoirs may not show any colouration but the toxins can still be present. In strong winds, sprays can be set up so that tiny droplets of contaminated water get taken in by birds causing damage to brain, kidneys and liver.

Tourism suffers if bodies of water become unusable for leisure pursuits such as swimming, fishing and sailing. Foul odours associated with algal deaths can make such areas unpleasant to visit.

WARNING
algal bloom

Lost revenue can be substantial, not just in the context of tourism but also to commercial fishing as well.

Use of pesticides

It is now common practice to grow crops as single cultures. An unfortunate consequence of this is that if a pest or disease can attack one of the plants, it can attack all of them. Crops can be infected by viruses, fungi or bacteria, or invaded by invertebrate animals such as insects, slugs and worms. Vertebrates, such as rats, can also become pests if they feed on grain and seeds. Growers may resort to using chemicals called **pesticides** to control pests. The crop yield is thereby increased, reducing costs and making harvesting easier and more efficient.

While strict legislation governs what chemicals can be used and how they are applied, it is still possible that some residual chemicals might be washed into waterways causing pollution. Chemicals can also find their way into food chains, thereby affecting balanced ecosystems. It is nearly impossible to find a pesticide that will target only the pest species without affecting other organisms as well. Past experiences have shown that wildlife and humans can be affected.

> **TOP TIP**
>
> One way of reducing the need for chemical pesticides is to use genetically modified crops which have pest resistance as part of their phenotype.

Various governmental agencies in Britain monitor the effect of prolonged use of pesticides and the effect on human health, even in small doses. For example, the chemical vinclozalin is used to control fungal infections such as rotting and moulds in vineyards and orchards on lettuces, beans, peas and onions. It is also often used by the people who manage golf courses. However, research has shown it can affect the reproductive hormone function in humans. Other anti-fungal agents have been shown to have similar effects on the human reproductive function by affecting sperm production and unborn babies.

> **TOP TIP**
>
> The use of pesticides has been linked with some health problems for humans such as cancer and infertility.

It is estimated that commercial growers now use only a fraction of the chemicals compared with 30 years ago. Some chemicals that were used freely, such as DDT, are no longer allowed. DDT is one of a number of pesticides that can persist in the environment and **bioaccumulate**. Since it is very soluble in fat, it can develop high levels in any food that is fatty, such as fish, meat and dairy products. It has been linked to cancer by affecting hormone production in men.

The use of biological control and genetically modified crops are alternatives to the use of pesticides.

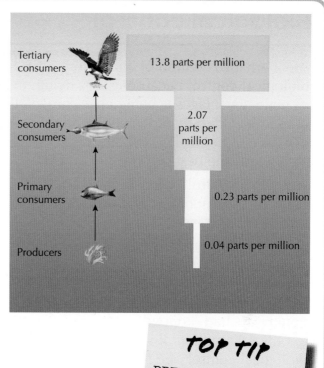

Tertiary consumers — 13.8 parts per million

2.07 parts per million

Secondary consumers

Primary consumers — 0.23 parts per million

0.04 parts per million

Producers

TOP TIP

DDT has been banned in the UK since the 1980s.

Biological control

An alternative method of controlling pests is being increasingly used. This is called **biological control** because it uses living organisms instead of chemicals. One of its main advantages is being environmentally friendly; it uses naturally occurring organisms rather than potentially harmful chemicals. The main form of this control involves importing known, target-specific predators from another region into the area where the pest organism is located.

Biological control has a number of advantages. It can provide a stable way of regulating the numbers of a pest, thereby improving crop production or saving a threatened species.

In the 1950s a virus that causes the disease **myxomatosis** in rabbits was deliberately introduced to control their numbers. The effects were dramatic and within two years, the population had dropped from about 600 million to only 100 million. It was the world's first biological control of a mammalian pest.

Most garden and greenhouse plants are attacked by **aphids**, insects that suck the sap from plants and stunt normal growth. Their numbers can be controlled by introducing a known predator, the ladybird, both in adult and larval forms. One ladybird can eat over 400 000 aphids in its lifetime! However, if the number of aphids is low, the ladybirds can experience a shortage of food and may migrate to other areas or, if they are inside a glasshouse, they will eventually starve to death.

TOP TIP

You need to know what biological control is and how it is used.

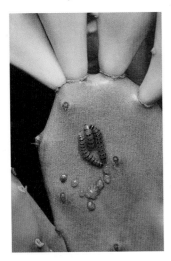

In some parts of Australia, the *Opuntia* cactus is a pest because of its rapid growth. In the mid-1920s, the only effective control was hazardous chemicals and, at the time, many farmers had to abandon large areas of their land. Now numbers of the cactus are controlled by a special moth that can detect particular chemicals found in this species. The larvae feed on the cactus, destroying it.

TOP TIP

The biological control organism does not eliminate the pest; it only controls its numbers to an acceptable level.

Biological control organisms work best if they possess the following features. They should:

- ideally be host specific or target only a very narrow range of hosts; organisms that feed on a wide variety of hosts will not reduce the population of the pest host if other food is available

- be able to tolerate possible new climatic conditions that may be warmer or cooler, drier or wetter than the control organism's normal environment

> ### TOP TIP
> Biological control is not a 'quick fix' solution because natural predators of a pest can take a long time to reduce the pest population to an acceptable level.

- kill the pest

- be present when the pest organism's life cycle first starts so that their numbers are checked quickly and effectively

> ### TOP TIP
> You might want to revise the topic of genetic engineering.

- be easy to work with in a laboratory

- have a good ability to locate the pest organisms

- have a high reproductive potential so that their numbers are adequate to suppress the target pest numbers

- not eliminate the target pest completely or it risks having not enough food to survive itself

- not harm the environment

- remain stable over a long period of time

- be easy to produce in large numbers.

An alternative strategy for controlling pests is to breed resistant varieties of crops. These are known as **genetically modified crops** because their genotypes have been changed to confer an in-built resistance to particular pests.

The technique for changing the genotype is genetic engineering. As well as engineering pest resistance, scientists have been able to produce plants with other desirable phenotypes, such as increased yields. Consequently a given area of farmland can be made to produce more food than previously. This helps to reduce the effects of intensive farming methods. Other enhanced phenotypes might allow plants to tolerate more extreme environmental conditions of temperature, lack of water and low light intensities.

The use of genetic engineering to modify crops has not been without its critics. There is a balance between the benefits and potential dangers that must be met before such a crop is allowed to be grown and ultimately end up in the human food chain.

Rapeseed plants can be engineered to resist a particular chemical used to control a pest infestation such as a weed. Farmers can spray the genetically modified crops with that chemical, eliminating the weed without damaging the plant. This means that there will be a much larger yield of the rapeseed plant, using a more environmentallyfriendly spray to control the pests and less of the spray. However, the genes conferring the resistant phenotype might find their way into other species because the pollen from rape seed plants can also pollinate weeds. In this way, the genes could confer resistance on those weeds which would then be resistant to the chemical.

Sweetcorn can be genetically engineered to produce a poison that kills insect pests. Therefore the farmer no longer needs to use any chemical sprays and the potential damage to the environment and other species is eliminated. The yield is also increased. However, this corn will continue to kill the insects for as long as it is growing, which gives rise to the possibility of a natural selection for insects that are resistant. In that case, the poison would no longer be effective. Other desirable insects, such as butterflies, might be killed as well as the target pest.

Scientists have been able to engineer rice so that it makes a chemical that humans can change into vitamin A. This rice therefore becomes a rich source of a vitamin, which is deficient in some diets, especially in underdeveloped areas. However, people in these areas may become overdependent on this rice. Since it is often linked with commercial enterprises, the producers ensure that the plants are sterile so that the growers cannot replant them and have to buy new stock each year.

Tomatoes previously had a relatively short shelf life because they produce a chemical that causes rotting. They can now be engineered to produce less of this chemical and so stay fresher for much longer.

Clearly there are a number of pros and cons to the development of genetically modified crops.

Advantages are:

- nutritional content can be enhanced
- taste is improved
- less chemicals need to be used
- money is saved
- the environment is not exposed to potentially harmful pesticides
- crops remain fresher for longer
- higher yields can be obtained
- crops can be transported for a longer time
- harvesting is easier if the crops are ready at the same time.

TOP TIP

You should discuss and investigate the ethical issues surrounding the use of genetically modified crops and food.

Concerns are:

- crops that are already genetically engineered to be resistant to antibiotics may pass on the genes conferring resistance to animals and people
- consumers are not always convinced the crops and their products are safe to eat
- other organisms might be affected if animals eat the genetically modified crops
- there might be no way to ensure control of the modified crops, which might spread from where they were initially grown
- genes that have been transferred from nuts into other organisms have caused major health concerns for people with nut allergies
- companies might end up with sole control of genetically modified crops
- the outcome is not always precise and predictable.

Quick Test

1.	State three factors which help keep the number of organisms in a population relatively constant.
2.	Explain how genetically modified food helps to feed the world's increasing population.
3.	Suggest three ways in which excess nutrients can enter waterways.
4.	State one advantage and one disadvantage of biological control.

Evolution of species

Mutations

Occasionally a change can occur in the genetic make-up of an organism. This is called a **mutation**. Mutations are random, spontaneous events and are in no way directed. They can occur anywhere in the genetic material of an organism's cells affecting that organism only. However, if mutations occur in the reproductive cells, the effect can be passed on to the next generation. Variations between individuals rely on mutations.

Mutations that have no effect on the organism are called **neutral**. Sometimes the effect may be positive and give some kind of advantage to survival; if negative it gives some kind of disadvantage to the survival of the organism.

TOP TIP

Mutations are spontaneous and are the only source of new alleles.

TOP TIP

If a mutation is advantageous, the organism is more likely to survive and pass on this mutated gene to the next generation.

The natural rate of mutations can be increased by environmental factors such as radiation and chemicals. Other agents which can increase mutation rates include:

- ultraviolet light
- X-rays
- gamma rays
- tars in cigarette smoke
- mustard gas.

Some viruses are known to induce mutations.

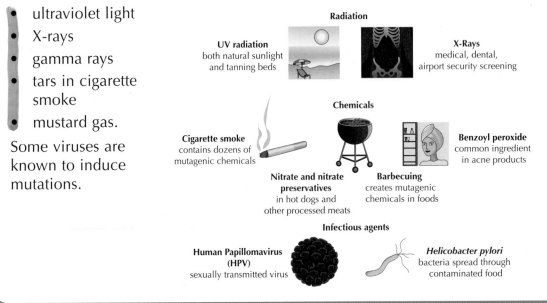

Radiation

UV radiation
both natural sunlight and tanning beds

X-Rays
medical, dental, airport security screening

Chemicals

Cigarette smoke
contains dozens of mutagenic chemicals

Nitrate and nitrate preservatives
in hot dogs and other processed meats

Barbecuing
creates mutagenic chemicals in foods

Benzoyl peroxide
common ingredient in acne products

Infectious agents

Human Papillomavirus (HPV)
sexually transmitted virus

Helicobacter pylori
bacteria spread through contaminated food

Variation within a population

Mutations are the basic cause of the variations that exist within and between species. New alleles produced by mutation can result in plants and animals becoming better adapted to their environment. Over a long period of time, often millions of years, inheritable variations allow organisms to change, or evolve, to meet different environmental conditions. This is why organisms often seem so well adapted to suit their particular habitat.

The desert rat is perfectly adapted for desert life. The colour of its fur blends in magnificently with its surroundings. Desert rats can survive without ever drinking natural water, obtaining the moisture they need from their food

TOP TIP

An adaptation is an inherited characteristic that makes an organism well suited to survival in its environment/niche.

during respiration. They have large eyes and good vision. Their hearing is so acute, they can pick up the sound of an approaching owl. Their muscular large back legs allow them to jump up to nearly three metres to escape predators!

Some orchids and moths have evolved together. The moths depend on the orchids to supply them with nectar, while the flowers depend on the moths to spread their pollen for reproduction. The moth has evolved a very long mouthpart that enables it to penetrate deeply into the orchid to obtain the nectar. As the moth sucks up the nectar, pollen sticks to its body. When the moth visits another orchid to feed, the pollen rubs off its body and pollinates the orchid.

The acacia tree has huge, hollow thorns that are ideal homes for ants. It also provides the ants with a ready source of food in the form of nectar and protein from specialised small structures that develop on the acacia leaves. The acacia benefits by having very aggressive guardians that ward off potential herbivores with their very strong sting.

TOP TIP

Investigate other examples of adaptations such as the effects of overuse of antibiotics.

Natural selection

The gradual development of organisms over very long periods of time is known as **evolution**. A theory of evolution was put forward by Charles Darwin in the 19th century. He made a number of observations, based on research and data collected over many years, then formed conclusions based on those observations.

Charles Darwin

Observations	Species produce more offspring than the environment can sustain
	Generally, populations of organisms tend to remain the same size over periods of time
Conclusion	Not all the offspring survive; many must die before being able to reproduce
Observation	Variations exist in populations of a species and these variations may be inheritable
Conclusions	The best adapted individuals in a population survive to reproduce passing on the favourable alleles that confer the selective advantage
	These alleles increase in frequency within the population
	Eventually, those organisms with the desirable features will dominate, perhaps to the point of eliminating other organisms that do not possess those same features
	With the passing of time, new species can evolve

The formation of a new species is called **speciation**.

One of the best examples of evolution in action is the peppered moth. It exists in two different forms; one that is brightly coloured and one that is black. This difference in phenotype is genetically based. Before the Industrial Revolution in the late 18th century the brightly coloured form was the most common. Its colour allowed it to blend in well against the barks of trees. The black-coloured one only appeared rarely as a mutated form and was spotted and eaten quickly by birds. However, due to the increase in air

TOP TIP

Evolution is usually a very gradual process which takes place over millions of years.

pollution during the Industrial Revolution, tree barks became increasingly darkened by soot giving the black moths the advantage over the brightly coloured forms. The black forms soon became the most common form because birds couldn't see them against the dark tree barks. By the middle of the 20th century, in polluted areas the black form dominated but in non-polluted areas the brightly coloured form was the most common.

TOP TIP

This is an example of 'high speed' evolution in action and can also be seen in the development of antibiotic-resistant strains of bacteria.

As air quality has improved dramatically in recent years the black forms are once again rarely found.

Speciation

The numbers and types of different species on Earth continue to change as some species become extinct and new species arise. Speciation occurs after part of a population becomes isolated by a barrier. These smaller groups are then somehow prevented from interbreeding. This is called **isolation**. Each sub-population will continue to breed independently of the others with variations arising due to mutations. Natural selection selects for different mutations in each sub-population due to different selective pressures. This means changes will increase until eventually the sub-populations are no longer able to interbreed, should the barriers be removed. At this point, the sub-populations would be termed new species.

TOP TIP

These changes take place over vast periods of time.

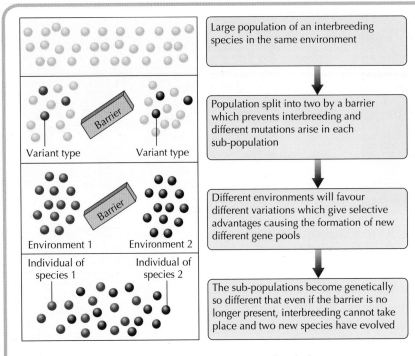

	Large population of an interbreeding species in the same environment
Variant type — Barrier — Variant type	Population split into two by a barrier which prevents interbreeding and different mutations arise in each sub-population
Environment 1 — Barrier — Environment 2	Different environments will favour different variations which give selective advantages causing the formation of new different gene pools
Individual of species 1 — Individual of species 2	The sub-populations become genetically so different that even if the barrier is no longer present, interbreeding cannot take place and two new species have evolved

There are several different ways in which barriers to gene transfer can be set up:

- Ecological
- Behavioural
- Geographical.

Ecological barriers include changes in environmental conditions such as increasing or decreasing temperature, humidity levels, pH and water availability, which can produce localised areas that are unsuitable for a population, thereby splitting it into one or more sub-populations.

Behavioural isolation is based on mating behaviour and associated rituals and signals. Two populations may be able to interbreed but are kept separated because of differences in their courtship behaviour.

Geographical barriers are physical in nature such as rivers, oceans, mountains and deserts, which prevent sub-populations exchanging genes.

The Isle of Arran is home to two of Scotland's rarest tree species: the Arran whitebeam and the Arran cut-leaved whitebeam. They are located in two glens that have different environmental conditions of altitude, which may have been an ecological barrier in the past, separating a single species into two sub-populations that eventually became two new species.

TOP TIP

Remember the sub-populations formed by barriers must acquire different mutations to allow different variations to arise.

Quick Test

1. 'A mutation is a change in an organism's genotype directed by a change in the environment.' Is this statement true or false? Explain your answer.

2. Describe three ways in which the desert rat minimises water loss.

3. The brightly coloured and black-coloured forms of the peppered moth are variations within the same species. How could you prove this?

Revision Questions

Section A

1. Which of the following is correct?

	Biotic factor	Abiotic factor
A	Length of day	Food supply
B	Altitude	Latitude
C	Competition	Predation
D	Disease	pH

2. Indicator species can give information about:

    A number of fish in a pond

    B number of prey in a woodland

    C strength of sunlight in a forest

    D the level of air pollution in a town.

3. In an ecological investigation samples should be:

    A random and few in number

    B representative and random

    C many in number and non-random

    D small in number and random.

4. The graph below represents the number of different fish species found in a variety of lakes of different pHs.

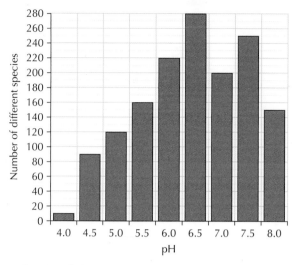

Which of the following statements is correct?

A The greatest number of different species is found at pH 4·0

B There are more different species at pH 8·0 than at pH 7·0

C With increasing pH there is a consistent increase in the number of different species

D The largest number of different species is found at pH 6·5.

5. The following statements refer to the events that might lead to an algal bloom in a river:

1. Algae near the surface of the river grow rapidly

2. Fertilisers are added to crops and run off due to excessive rain

3. Nitrate levels in the river rise

4. Algae form a carpet over the surface of the river.

Which is the correct order of events?

A 1, 2, 3, 4

B 4, 3, 2, 1

C 2, 1, 3, 4

D 2, 3, 1, 4

Section B

1. (a) The diagram below shows parts of a process that takes place in plants.

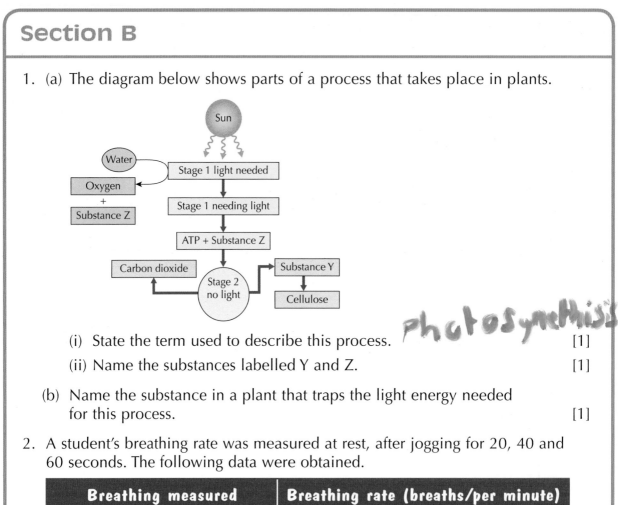

      (i) State the term used to describe this process. *Photosynthesis* [1]

      (ii) Name the substances labelled Y and Z. [1]

  (b) Name the substance in a plant that traps the light energy needed for this process. [1]

2. A student's breathing rate was measured at rest, after jogging for 20, 40 and 60 seconds. The following data were obtained.

Breathing measured	Breathing rate (breaths/per minute)
At rest	15
After jogging for 20 seconds	20
After jogging for 40 seconds	30
After jogging for 60 seconds	40

  (a) Plot these results as a **bar chart**. [3]

  (b) Calculate the percentage increase in the breathing rate from at rest to after jogging for 60 seconds. [1]

  (c) Calculate as a **simple whole number** ratio, the breathing rates at each of the four measured periods. [1]

3. State the term used to describe the use of a living organism to control a pest. [1]

4. Name three variables that can affect an organism's niche. [1]

5. The following is a simple pyramid of energy:

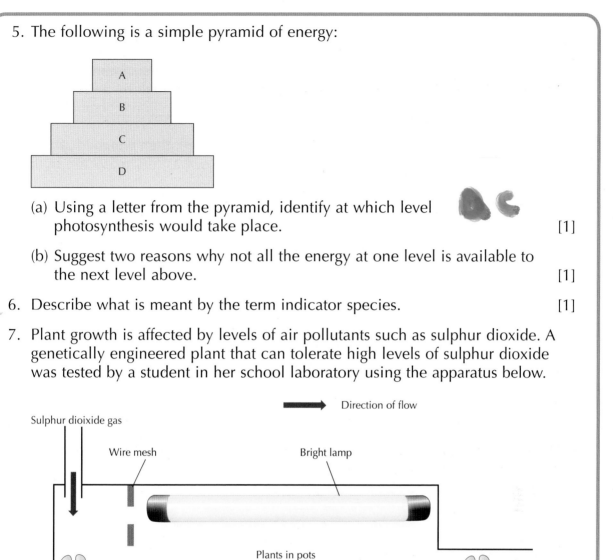

(a) Using a letter from the pyramid, identify at which level
    photosynthesis would take place.   [1]

(b) Suggest two reasons why not all the energy at one level is available to
    the next level above.   [1]

6. Describe what is meant by the term indicator species.   [1]

7. Plant growth is affected by levels of air pollutants such as sulphur dioxide. A
   genetically engineered plant that can tolerate high levels of sulphur dioxide
   was tested by a student in her school laboratory using the apparatus below.

Five samples of the genetically engineered plants were grown in pots and then placed in the chamber. Sulphur dioxide gas was mixed with normal air using a fan to give a known concentration well above the usual atmospheric concentrations of the gas and the mixture was drawn through by a second fan. A bright lamp was kept on throughout the experiment. A similar chamber, with another five genetically engineered plants, was set up as a control using only normal air. After six weeks, the student collected, dried, then weighed all the plant material from both chambers. The results, shown as 'dry mass' are in the following table.

Chamber	Dry mass of each plant (g)					Average dry mass (g)
Experimental with sulphur dioxide gas	12	10	11	12	10	
Control with normal air	18	20	18	21	23	

(a) Copy and complete the table by calculating the average dry mass for each chamber. [1]

(b) In the experimental set-up, explain the purpose of each of the following:

   (i)   Wire mesh

   (ii) Bright lamp

   (iii) Fan A

   (iv) Fan B [4]

(c) Describe two precautions the student would have to take when measuring the dry mass of each plant. [2]

(d) Give three variables that should be kept the same throughout the investigation. [3]

(e) Give an example of a flaw in the design of this investigation and explain how you would attempt to correct it. [2]

(f) Describe how the student could improve the reliability of the results. [1]

(g) The student concluded that the decrease in the average growth rate of the plants in the sulphur dioxide-enriched atmosphere was due to an inability of the plants to photosynthesise properly.

   Was this a valid conclusion? Explain your answer. [2]

Glossary

abiotic factor
any condition affecting the environment that results from non-living sources

absorption
process by which small, soluble molecules are taken up by cells

active site
area on an enzyme's surface that binds with a specific substrate

active transport
energy-demanding process in cells that moves substances against a concentration gradient

adenine
one of four bases found in DNA

aerobic respiration
type of respiration requiring oxygen in which substrates such as glucose are completely broken down to water and carbon dioxide to release large amounts of energy

albinism
genetically determined condition that results in none of the normal pigment found in hair, skin and eyes

alga
simple one-celled or many-celled plant found mostly in water which uses sunlight to produce its own food

algal bloom
excess algae often associated with an increase in the nutrient level in a body of water

allele
different forms of a gene

alveolus
commonly called an air sac, blind-ending, thin-walled sac where gas exchange takes place in the lungs

amino acid
basic building block of a protein

amylase
enzyme that breaks starch down into the sugar maltose

anther
male structure in a flowering plant which produces pollen

antibody
large protein molecule produced in response to invasion by a foreign agent and capable of rendering it harmless

aorta
large artery taking blood rich in oxygen from the heart

aphid
sap-sucking insect that stunts the growth of plants it infects

artery
vessel carrying blood from the heart

atrium
upper thin-walled chamber of the heart that receives blood from the body or the lungs

bacterial cell
microscopic one-celled organism with no nucleus but with a definite cell wall and plasmids

base
nitrogen-containing chemical such as adenine, thymine, cytosine or guanine

biconcave
being curved inwards on both sides

biodiversity
all the different species that live in an environment

bioaccumulation
build up of pesticides accumulating in the bodies of organisms over time

biological control
method of controlling pests using naturally occurring living organisms to regulate the size of the pest population

Glossary

biosphere
part of the Earth's surface and atmosphere within which life can exist

biotic factor
anything that affects the environment as a result of the activities of living things

brain
complex of nerve cells found inside the skull where all the higher level activities take place

capillary
smallest diameter blood vessel whose walls are only one cell thick and across which an exchange of gases, nutrients and wastes takes place

carbohydrate
chemical containing the elements carbon, hydrogen and oxygen

carbon fixation
combination, by photosynthetic plants, of the gas carbon dioxide with hydrogen to produce glucose

carnivore
animal which feeds only on other animals

catalyst
chemical that can speed up a reaction

cell membrane
outer covering of cells that regulates what can enter or leave

cell wall
relatively thick layer found on the outside of plant, fungal and bacterial cells. In each case, it is chemically different but functions to give the cell shape and helps protect internal cell structures

cellulose
main structural chemical that makes up plant cell walls

central nervous system (CNS)
consists of brain and spinal cord

cerebellum
part of the brain associated with balance and co-ordinating voluntary muscle activity

cerebral hemisphere
one of a pair of large lobes that form the cerebrum

cerebrum
largest part of the brain occupying space from above eyes to back of the head

cervix
muscular ring at the neck of the uterus leading into the vagina

chamber
in the heart, one of the four cavities that collect and discharge blood

chlorophyll
green pigment found in chloroplasts capable of trapping light energy

chloroplast
structure found in green plant cells that contains the pigment chlorophyll and where photosynthesis takes place

chromatid
one of two identical strands forming a chromosome

chromosome
thread-like structure composed of deoxyribonucleic acid found in the nucleus and carrying genetic instructions

chromosome complement
total number and types of chromosomes found in the nucleus of a cell

circulatory system
collective name for the blood, vessels and heart

collagen
important chemical found in bone and skin that gives strength

community
collection of animals and plants living together in a particular habitat

companion cell
nucleus helps regulate activity of the sieve tube cell

complementary base pairing
linking of bases by hydrogen bonding in specific combinations

concentration gradient
difference in the concentration of substances from one area to another

consumer
animal which feeds on other organisms to obtain energy

continuous variation
continual spread across a sample of a population for variables that are often controlled by a number of factors

cuticle
continuous waxy layer that covers the surfaces exposed to the air

cytoplasm
watery substance found inside cells where all the chemical reactions of the cell take place

cytosine
one of four bases found in DNA

decomposer
any living thing that can break down dead material to allow nutrients to be recycled in ecosystems

degradation
breakdown of large molecules into smaller ones

denaturation
irreversible change in a protein, caused by changes in pH or temperature, that renders the activity of an enzyme useless

deoxyribonucleic acid (DNA)
complex, helically-shaped molecule of heredity within which are encoded instructions for constructing, controlling and reproducing cells by determining the synthesis of proteins

deoxyribose
5-carbon sugar found in DNA

diaphragm
strong sheet of muscle separating the chest cavity from the lower gut cavity and is important in breathing

diffusion
sometimes referred to as passive transport, is the movement of substances from an area of high concentration to an area of low concentration without the use of energy

diploid number
total number of chromosomes in the nucleus

discrete variation
differences in a particular feature fall into distinct categories and cannot be easily measured

dominant allele
form of a gene that masks the effect of the recessive form of the allele and produces a dominant phenotype

double helix
the characteristic shape of the DNA molecule which consists of two strands, each of which turns regularly about itself to form a cylindrical shape held together by weak hydrogen bonds

ecological barrier
means by which a species becomes divided into two or more sub-groups based on some change in the environment

ecosystem
all the organisms (the community) living in a particular habitat and the non-living components with which the organisms interact

effector
structure that brings about an action as a result of an input from a nerve pathway

egg
female sex cell

embryo
in animals developmental stage from foetal stage up to time of birth

endocrine gland
collection of cells that produce and release chemicals directly into the bloodstream

endocrine system
collection of glands that release chemicals directly into the bloodstream

enzyme
protein that has the ability to cause a reaction in a living cell to take place quickly when it would otherwise take place slowly or not at all

epidermis
in a multicellular organism the outer layer of cells, usually one cell thick in plants

equator
plane of cell along which chromosomes line up during mitosis

Glossary

ethanol
an alcohol produced during anaerobic respiration by plant cells

evolution
process by which living things have gradually changed over a very long period of time to become better suited to survive and reproduce in their environment

family tree
pictorial representation of a family's inheritance patterns over a number of generations

fat
chemical, usually solid at room temperature, used as an energy-store

fatty acid
chemical that, when combined with glycerol, forms fat molecules

fermentation
type of anaerobic respiration found in plant and yeast cells that results in the production of ethanol, carbon dioxide and small quantities of ATP

fertiliser
natural or synthetically produced chemical added to soil to enhance its propeties in some way

fertilisation
fusion of male and female gametes to form a zygote

first filial generation
offspring produced as a result of a parental cross

foetus
young animal in its early stages of development and still within the mother's uterus

food chain
simple, linear representation of the feeding relationship of animals and plants

food web
complex set of feeding relationships between animals and plants consisting of many linked food chains

fungus
organism that has no chlorophyll, has a cell wall made of chitin and often feeds on dead animal or plant materials

gamete
sex cell that possesses half the diploid number of chromosomes

gene
basic unit of heredity which corresponds to a length of DNA

genetic engineering
term for different techniques to deliberately alter the DNA of a cell by inserting part or all of the genetic material from another cell which may or may not be from the same species or organism

genetically modified crops
plants whose DNA has been changed by genetic engineering

genetically modified
description of a cell's genetic material after it has been changed by genetic engineering

genotype
the combination of the alleles of a gene or genes

geographical barrier
means by which a species becomes divided into two or more sub-groups based on some major physical obstacle that prevents the sub-groups from continuing to interbreed

gland
organ producing a chemical that brings about a response in any body part which is sensitive to that chemical

glucagon
hormone produced by the pancreas that causes glycogen to be converted to glucose

glucose
simple 6-carbon sugar that is a product of photosynthesis and is used up in respiration

glycerol
a basic component of fat molecules

glycogen
main sugar made up of many glucose molecules and stored in the liver

guanine
one of four bases found in DNA

guard cell
specialised cell that surrounds a stoma and regulates the size of the opening

habitat
general term for the place in an environment where an organism lives

haemoglobin
protein that combines loosely with oxygen in the lungs and then offloads this in respiring tissues

haploid number
number of chromosomes present in a sex cell

heart
muscular pump situated in the chest cavity behind the breastbone

herbivore
animal which feeds only on plants

heterozygote
an individual with different alleles of the same gene

homozygote
an individual with identical alleles of the same gene

hormone
chemical produced by one part of a plant or animal and transported to target areas to affect function and/or structure

immune system
collective name for the cells, tissues and organs which bring about a response to a pathogen and may allow the development of long-term resistance

impulse
of nerves, the message conducted along a nerve

indicator species
organism whose presence or absence indicates the condition of a habitat

inheritance
how characteristics in living things are passed from one generation to another

insulin
hormone that regulates the blood glucose levels by converting glucose to glycogen

inter neuron
nerve cell carrying information from sensory neurone to motor neurone

interspecific competition
competition between organisms of different species for the same resources

intraspecific competition
competition within organisms of the same species for the same resources

isolation
when a population is split into two or more smaller groups that are prevented from genetic exchange

key
system for identification based on observable features, present or absent, of organisms

lacteal
central structure found in each villus that absorbs digested fats

lactate
compound formed in animal cells as an endproduct of anaerobic respiration during activities which have a high oxygendemand

leaching
process by which dissolved substances, such as phosphates and nitrates in the soil, are washed out by rainwater

lichen
fungi and algae growing together with fungus making up almost 90 % of the mass of the combination

lignin
substance found in the xylem of some plant cell walls that it stiffens, helping to stop infection and decay and making the xylem strong

limiting factor
variable that, when increased or decreased, speeds up or slows down a reaction or process

lymphocyte
white blood cell involved in defence and may be capable of producing antibodies

medulla
part of brain which connects with spinal cord and controls activities that animals are not consciously aware of

Glossary

messenger ribonucleic acid (mRNA)
chemical that is important in the manufacture of proteins carrying information from the DNA in the nucleus to the ribosomes in the cytoplasm

micron
unit of measurement for cells where 1 mm is equal to 1000 microns

mitochondrion
cylindrically-shaped structure found in varying numbers in the cytoplasm of cells that is the site of aerobic respiration producing adenosine triphosphate (ATP)

mitosis
type of nuclear division that results in the formation of two new cells that share the same genetic instructions as each other and the original cell from which they arose

motor neuron
nerve cell carrying information towards an effector

multicellular
describing an organism whose body is made up of many cells

mutation
change in the genetic make-up of a cell that can result in an altered phenotype, producing a new allele if a gene is affected, or a change in the number of the chromosomes

myxomatosis
deadly viral infection of rabbits

natural selection
mechanism by which gradual evolutionary changes take place

nervous system
collection of structures that allows a multicellular animal to coordinate its activities very rapidly

neuron
nerve cell

neutral mutation
change in the genetic make-up of a cell that has no effect on the organism

niche
role played by a particular organism in the environment that is usually a function of the food eaten and a range of variables tolerated

nucleus
controls all the activities of a cell and contains the genetic material

oesophagus
tube connecting mouth to stomach along which food passes

oil
fat which is liquid at room temperature

omnivore
animal which feeds on both plants and animals

optimum
value of a factor, such as pH or temperature, at which an enzyme works best

organ
functional unit in a multicellular organism which carries out a specific job

osmosis
movement of water from an area of high water concentration to an area of low water concentration across a selectively permeable membrane

ovary
organ in which female sex cells are produced in animals and plants

oviduct
tube that carries egg from ovary towards the uterus

ovule
contains the female gamete in plants

ovum
female gamete found in animals

oxyhaemoglobin
formed when haemoglobin combines with oxygen

palisade mesophyll
cell that forms a layer between the upper and lower epidermis of a leaf, shaped tall and columnar, and where photosynthesis mainly takes place in a leaf

pancreas
organ associated with the digestive system producing important enzymes and the hormones insulin and glucagon

parental generation
two parents that produce offspring

pathogen
agent such as a virus or bacterial cell which is capable of producing disease

passive transport
movement of substances from an area of high concentration to an area of low concentration without the use of energy

pesticide
chemical used to kill pests

phagocyte
white blood cell which can engulf and destroy pathogens

phagocytosis
process of ingesting and digesting a pathogen by a phagocyte

phenotype
expression of the genes possessed by an individual that is usually a combination of the effects of the genes and the environment

phloem
plant structure that moves food material made in the leaves by photosynthesis to other parts of the plant

phospholipid
fat molecule with a phosphate group attached

photosynthesis
process by which green plant cells use the energy of the sun to combine carbon dioxide and water to form carbohydrate

pigment
coloured compound produced by a living cell

pitfall trap
usually a jar or tin can sunk into the ground into which animals fall and are trapped

plasma
straw-coloured liquid part of blood in which cells are suspended

plasmid
small circular piece of genetic material commonly found in bacteria and usually made of deoxyribonucleic acid (DNA) that can reproduce independently of the main genetic material

plasmolysed
of plant cells, the condition of excessive water loss

pollen grain
contains the male gamete in plants

pollen tube
narrow structure down which the male gamete travels towards the ovule

polygenic
description of an inherited feature controlled by more than one gene

pooter
chamber with two tubes attached, one of which is put into the mouth and sucked through while the other draws a small invertebrates into the chamber for later examination

population
group of living things that belong to the same species and live in the same area of the environment

predator
animal which feeds on another animal as its food source

prey
animal which is the food source of another animal

producer
green plant at start of a food chain which photosynthesises to make its own food

product
end result of an enzyme-catalysed reaction

pulmonary artery
vessel carrying blood low in oxygen from the right ventricle to the lungs

pulmonary vein
vessel carrying blood rich in oxygen from the lungs to the left atrium

Glossary

Punnett square
simple table to show all possible results from a genetic cross

pyramid of energy
graphical representation showing the energy stored at each feeding level in an ecosystem

pyramid of numbers
graphical representation of the total number of living things at each feeding level in an ecosystem

pyruvate
important 3-carbon molecule that is an intermediate in respiration

receptor
cell or group of cells that responds to a specific stimulus

receptor
group of molecules on a cell membrane that fits another complementary molecule so that when the two link up, a change in cell function takes place

recessive allele
form of a gene that needs another similar allele for the recessive phenotype to be expressed

red blood cell
cell that has no nucleus and contains the protein haemoglobin to carry oxygen

reflex action
action not usually requiring the brain to be involved and is therefore unconscious

reflex arc
simple nerve pathway connecting a receptor and effector resulting in a specific response to a specific stimulus

reproductive barrier
means by which a species becomes divided into two or more sub-groups because they are incompatible in some way or the offspring produced are not fertile

respiration
process by which energy-rich molecules are progressively broken down by enzymes to form adenosine triphosphate

ribosome
small particle which is the site of protein synthesis in a cell

root hair cell
cell found in the root of a plant that has a very large surface area for absorbing water and dissolved solutes

sampling
technique of counting small numbers of a variable as a way of representing the actual number

second filial generation
offspring produced as a result of crossing two members of the first filial generation

selectively permeable
describes how a cell membrane exerts control on the substances that can pass across it

sensory neuron
nerve cell carrying information from a receptor to the central nervous system

sensory receptor
specialised cell or cells that detect stimuli and relay these to the central nervous system

sex cell
gamete or haploid reproductive cell

sieve plate
series of holes found at the end of a sieve tube cell which allows cytoplasmic connections between cells above and below

sieve tube
series of sieve cells which lie end to end to form a hollow canal

small intestine
narrow tube about seven metres long starting at the stomach where digestion is completed and absorption takes place

specialisation
when describing a cell, the state of being dedicated to one particular function

speciation
formation of two or more groups of organisms that can no longer interbreed to form fertile offspring

species
group of individuals which can breed together to produce offspring that themselves can reproduce

sperm duct
tube carrying sperms from testes to outside via the penis

sperm
male sex cell which contains half the normal number of chromosomes

spinal column
series of small bones stacked on top of each other that form a tube within which the spinal cord is protected

spinal cord
thick cable-like structure that carries information to and from the brain

spindle
network of fibres that appear during mitosis and move chromosomes within the cell

spongy mesophyll layer
cells that are irregularly shaped found between the upper and lower epidermis creating a large surface area and many air spaces within a leaf

starch
molecule made up of many glucose units joined together

stem cell
cell that is capable of growing into many different types of cell found in the adult animal or plant

stimulus
energy event in the environment or inside an animal's body that can be detected and potentially produce a response

stoma
small opening on the surface of leaves and stems that allows exchange of materials between the environment and the plant

structural protein
long chain of amino acids used as part of the fabric of a cell such as part of the membrane

substrate
chemical on which an enzyme acts

synapse
microscopic space between neurons

synthesis
building up of large molecules from smaller ones

system
collection of organs dedicated to a particular function

target tissue
collection of cells that are sensitive to a hormone

territory
specific area in an environment which is inhabited by one species that will not allow members of the same species to share it and that is defended by the resident member

testis
organ in male animal where sperm are produced

thymine
one of four bases found in DNA

tissue
collection of similar cells that perform a specific task

transpiration
evaporative loss of water through the surfaces of a land plant

transpiration stream
movement of water through a plant from the roots to the leaves

turgid
condition of a plant cell that is full of water

ultrastructure
fine detail of a cell as revealed by the electron microscope

uterus
muscular organ in which the embryo develops

vacuole
membrane-bound sac found in plant cells containing a watery solution giving support

Glossary

vagina
muscular tube that receives the erect penis during sexual reproduction and through which the baby passes during birth

valve
structure that allows movement of contents of a vessel or blood in the heart to flow in one direction only

variable
quantity that can continually increase or decrease

variation
difference which exists between living things that may be a function of the genetic make-up or the environment or a combination of both

vein
vessel carrying blood to the heart

vena cava
large vein returning blood low in oxygen to the right atrium of the heart

ventricle
thick-walled chamber of the heart that sends blood to the lungs or to the body

vessel
in mammals any tube that carries fluid such as blood

villus
finger-like projection lining the small intestine increasing the surface area available for absorption

voluntary muscle
muscle whose activity is under conscious control

white blood cell
common name for blood cell which is involved in defence

xylem
group of plant cells that transport water and dissolved solutes from the roots to the leaves

yeast
general term for a fungus that exists as a single-celled organism

zygote
fertilised egg at the one-cell stage that contains a full set of chromosomes

Leckie
e education publisher
or Scotland

National 5
BIOLOGY

or SQA 2019 and beyond

Practice Papers

Billy Dickson and
Graham Moffat

Revision advice

Design of the papers

Each paper has been carefully assembled to be very similar to a typical National 5 question paper. Each paper has 100 marks and is divided into two sections.

- **Section 1** – objective test, which contains 25 multiple-choice items worth 1 mark each, totalling 25 marks.

- **Section 2** – paper 2, which contains structured and extended-response questions worth 1 to 4 marks each, totalling 75 marks.

In each paper, the marks are distributed evenly across all three component areas of the course, and the majority of the marks are for the demonstration and application of knowledge. The other marks are for the application of skills of scientific inquiry. We have included features of the national papers such as including a scientific literacy question, offering choice in some questions and building in opportunities for candidates to suggest adjustments to investigation and experimental designs.

Most questions in each paper are set at the standard of Grade C, but there are also more difficult questions set at the standard for Grade A. We have attempted to construct each paper to represent the typical range of demand in a National 5 Biology paper.

Using the papers

We recommend working between attempting the questions and studying their expected answers.

You will need a **pen**, a **sharp pencil**, **a clear plastic ruler** and a **calculator** for the best results. A couple of different **coloured highlighters** could also be handy.

Expected answers

The expected answers online at www.leckieandleckie.co.uk give national standard answers but, occasionally, there may be other acceptable answers. The answers have Top Tips provided alongside but don't feel you need to use them all!

The Top Tips include hints on the biology itself as well as some memory ideas, a focus on traditionally difficult areas, advice on the wording of answers and notes of commonly made errors.

Grading

The two papers are designed to be equally demanding and to reflect the national standard of a typical SQA paper. Each paper has 100 marks – if you score 50 marks, that's a C pass. You will need about 60 marks for a B pass and about 70 marks for an A. These figures are a rough guide only.

Timing

If you are attempting a full paper, limit yourself to **2 hours and 30 minutes** to complete it. Get someone to time you! We recommend no more than 30 minutes for **Section 1** and the remainder of the time for **Section 2**.

If you are tackling blocks of questions, give yourself about a minute and a half per mark; for example, 10 marks of questions should take no longer than 15 minutes.

Good luck!

Practice paper A

SECTION 1 ANSWER GRID

Mark the correct answer as shown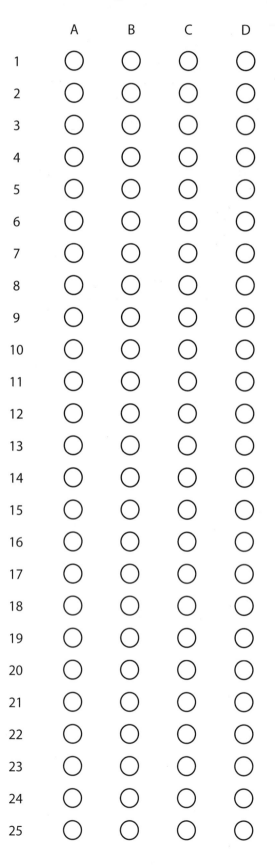

	A	B	C	D
1	○	○	○	○
2	○	○	○	○
3	○	○	○	○
4	○	○	○	○
5	○	○	○	○
6	○	○	○	○
7	○	○	○	○
8	○	○	○	○
9	○	○	○	○
10	○	○	○	○
11	○	○	○	○
12	○	○	○	○
13	○	○	○	○
14	○	○	○	○
15	○	○	○	○
16	○	○	○	○
17	○	○	○	○
18	○	○	○	○
19	○	○	○	○
20	○	○	○	○
21	○	○	○	○
22	○	○	○	○
23	○	○	○	○
24	○	○	○	○
25	○	○	○	○

N5 Biology

Practice Papers for SQA Exams

Practice Paper A
Section 1

Fill in these boxes and read what is printed below.

Full name of centre

Town

Forename(s)

Surname

Try to answer ALL of the questions in the time allowed.

You have 2 hours and 30 minutes to complete this paper.

Write your answers in the spaces provided, including all of your working.

×Leckie
the education publisher
for Scotland

SECTION 1 – 25 marks
Attempt ALL questions

1. The diagram below shows some structures present in a mesophyll cell from a green plant.

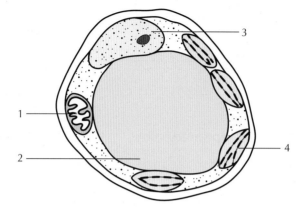

Which line in the table below identifies correctly the structures in the cell that carry out photosynthesis and contain genetic information?

	Carry out photosynthesis	Contain genetic information
A	1	2
B	4	3
C	1	3
D	4	2

2. The histogram below shows the number of cells of different lengths in a sample of onion epidermis.

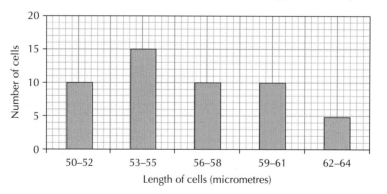

What percentage of the cells in the sample have a length greater than 58 micrometres?

A 15%

B 25%

C 30%

D 50%

3. Which line in the table below compares cell walls of plant and fungal cells?

	Chemical composition	*Structure*
A	same	same
B	different	different
C	same	different
D	different	same

4. The diagram below represents a bacterial cell as viewed under a microscope set to magnify 500 times.

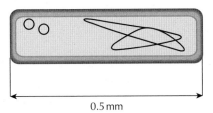

0.5 mm

How many cells of this size would fit end to end into a millimetre?

A 10

B 100

C 500

D 1000

5. 50 mm strips of potato tissue were placed into each of three sucrose solutions P, Q and R of different concentrations and left at room temperature. After 1 hour the strips of tissue were re-measured and the results are shown in the table below.

Sucrose solution	Length of potato tissue strip after 1 hour (mm)
P	50
Q	47
R	52

Which of the following conclusions based on these results is valid?

A Solution P had a lower concentration of sucrose than the potato cell sap

B Solution Q had a higher concentration of sucrose than the potato cell sap

C Solution R had a higher concentration of sucrose than the potato cell sap

D Solutions P, Q and R had the same concentration as the potato cell sap.

6. In active transport, molecules are moved by membrane

A proteins against the concentration gradient

B lipids down the concentration gradient

C lipids against the concentration gradient

D proteins down the concentration gradient.

7. The diagram below shows a stage in mitosis in a plant cell.

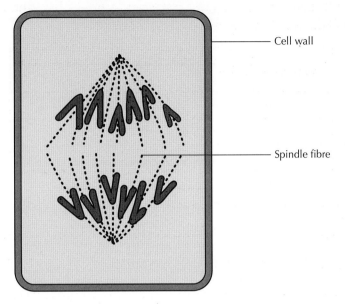

Which of the following best describes the chromosomes at the stage of mitosis shown?
The chromosomes have

A become visible as pairs of identical chromatids

B aligned at the equator of the spindle

C gathered at opposite poles of the spindle

D been pulled apart by spindle fibres.

8. The diagram below represents a short piece of a DNA molecule.

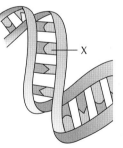

Which part of the DNA molecule is shown at X?

A Sugar

B Base

C Gene

D Amino acid.

9. The diagram below shows a genetically modified bacterial cell that contains a human gene.

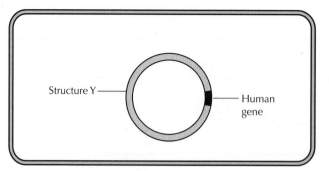

Structure Y, which contains the human gene, is

A the nucleus

B a chromosome

C a ribosome

D a plasmid.

10. A group of similar cells working together to perform the same function is called

A an organism

B a system

C an organ

D a tissue.

11. Which of the following statements is **false** in relation to stem cells?

Stem cells

A are found in animal embryos

B can undergo cell division

C develop into gametes

D can self-renew.

12. The diagram below shows a vertical section through the human brain.

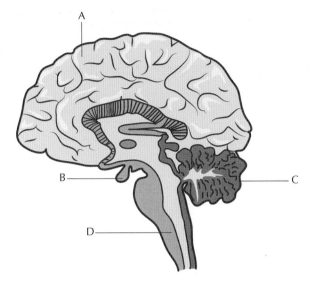

Which letter indicates the site of memory storage and reasoning?

13. Which organ contains target tissues that respond to insulin?

A Small intestine   C Liver

B Pancreas   D Brain.

14. Which line in the table below shows correctly the chromosome complements of the mammalian cells listed?

	Mammalian cell		
	muscle cell	gamete	zygote
A	diploid	haploid	haploid
B	diploid	haploid	diploid
C	haploid	diploid	diploid
D	haploid	diploid	haploid

15. The cardiac output from the heart is calculated using the equation shown below.

cardiac output (litres per min) = volume of blood pumped per beat (cm³) × heart rate (beats per minute)

A hospital patient had a heart rate of 80 beats per minute and a cardiac output of 4 litres per minute.

What is the volume of blood pumped per beat?

A 5 cm^3 C 50 cm^3

B 20 cm^3 D 320 cm^3

Questions 16 and 17 refer to the following information.

A weight potometer was set up to compare the transpiration rates of a plant in different sets of environmental conditions.

The graph below shows the results of two experiments in which the environmental conditions were altered.

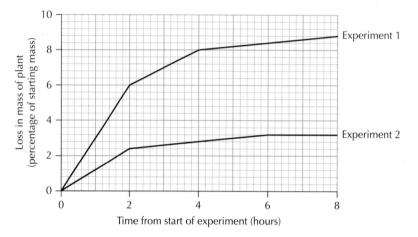

16. Which line in the table shows the possible conditions in Experiments 1 and 2 assuming all other conditions were kept constant?

	Experiment 1		Experiment 2	
	temperature	humidity	temperature	humidity
A	high	low	low	high
B	high	low	high	low
C	low	high	high	low
D	low	high	low	high

17. In Experiment 1, the plant had a mass of 600g at the start of the experiment.

What was its mass after 4 hours?

A 547.2g

B 552.0g

C 583.2g

D 592.0g

18. The total variety of all living organisms on Earth is described as its

A habitat

B biodiversity

C ecosystem

D population.

19. Which of the following statements is **true**?

The community of a Scottish moorland ecosystem consists of all the

A plant species present

B plant species present and the non-living environment

C plant and animal species present and the non-living environment

D plant and animal species present.

20. Which of the following factors are **both** biotic?

A Predation and temperature

B Temperature and pH

C pH and grazing

D Grazing and predation.

21. In the food chain below the plant plankton contains 100 000 units of energy from photosynthesis.

plant plankton $\rightarrow$ **animal plankton** $\rightarrow$ **small fish** $\rightarrow$ **predatory fish**

If 90% of the energy available at a food chain level is lost between levels, how many units of energy will be found in the predatory fish in the food chain above?

A 10 000

B 1000

C 100

D 10.

22. Which of these substances is absorbed from soil by plants and used in the synthesis of proteins?

A Amino acids

B Nitrates

C Sugars

D Water.

Questions 23 and 24 refer to the following information.

During a survey of the distribution of limpets on a rocky seashore, a number of quadrat samples were taken along a transect line between the high and low tide marks.

The numbers in the diagram below indicate the numbers of limpets in each quadrat.

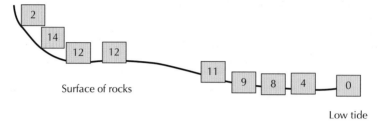

23. What is the average number of limpets per quadrat?

A 8

B 9

C 11

D 12

24. Which of the following is a precaution needed to make the results of the survey more valid?

A Place quadrats randomly

B Use exactly ten quadrats

C Place quadrats where limpets occurred

D Repeat the quadrat sampling several times.

25. Which of the following is a source of variation in a species of mammal?

A Isolation

B Natural selection

C Mutation

D Adaptation.

N5 Biology

Practice Papers for SQA Exams

Practice Paper A
Section 2

Fill in these boxes and read what is printed below.

Full name of centre

Town

Forename(s)

Surname

Try to answer ALL of the questions in the time allowed.

You have 2 hours and 30 minutes to complete this paper.

Write your answers in the spaces provided, including all of your working.

SECTION 2 – 75 marks
Attempt ALL questions

1. The diagram represents molecules present in a magnified fragment of cell membrane.

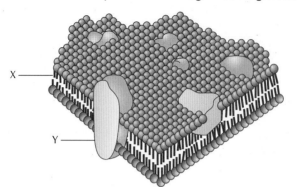

(a) Name molecules X and Y.

X _____

Y _____

2

(b) Complete the following sentences by <u>underlining</u> the correct options in each choice bracket.

The cell membrane is $\begin{Bmatrix} \text{selectively} \\ \text{fully} \end{Bmatrix}$ permeable and transports water in and out of the cell by osmosis.

Osmosis occurs $\begin{Bmatrix} \text{down} \\ \text{against} \end{Bmatrix}$ the concentration gradient and

and $\begin{Bmatrix} \text{requires} \\ \text{does not require} \end{Bmatrix}$ energy.

2

(c) The diagram below shows a cell from a piece of plant tissue.

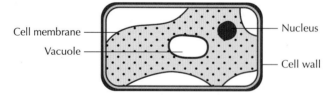

(i) Describe how a piece of plant tissue could be treated so that its cells appeared as shown in the diagram.

_____

_____

1

(ii) Give the term applied to cells that appear as shown in the diagram.

_____

1

Total marks 6

2. The diagrams below represent stages in a synthesis (building up) reaction catalysed by a human enzyme molecule at 37 °C.

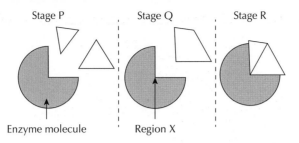

Stage P   Stage Q   Stage R

Enzyme molecule   Region X

(a) Complete the flow chart below by adding letters to show the correct order of these stages as they would occur during the reaction.

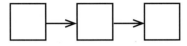

1

(b) Identify the part of the enzyme molecule labelled Region X in the diagram.

_____

1

(c) Give the term which would be used to describe Stage R.

_____

1

(d) Explain why the cellular reaction above would **not** occur if the temperature were increased to 60 °C.

_____

_____

2

(e) Apart from temperature, give **one** other factor which could affect the rate of an enzyme- catalysed reaction.

_____

1

Total marks **6**

3. An investigation was carried out on the effect of temperature on the rate of fermentation in yeast.

Apparatus as shown in the diagram below was set up, and the number of bubbles of gas produced by the yeast per minute was counted at various temperatures, as shown in the table.

Apparatus

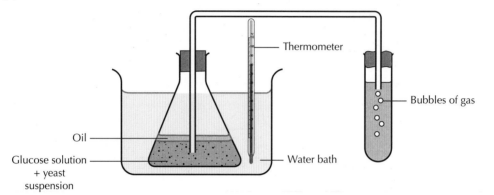

Temperature (°C)	Bubbles of gas produced per minute
10	30
15	50
20	80
25	110
30	120

MARKS
Do not write in this margin

(a) On the grid provided below, complete the line graph to show temperature against number of bubbles of gas produced per minute.

(A spare grid, if required, can be found at the end of the practice paper.)

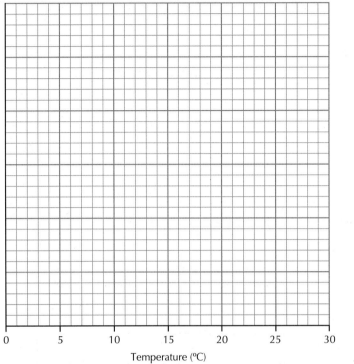

Temperature (°C)

2

(b) Identify the gas produced during fermentation.

_____

1

(c) Suggest how the investigation could be improved to give more accurate results.

_____

_____

1

(d) Predict how the results would be different if the investigation were repeated at 5 °C. Explain your answer.

Prediction _____

1

Explanation

_____

_____

1

Total marks 6

4. In an investigation into beer production, a fermenter containing a glucose solution and a yeast culture was set up. The concentrations of glucose and ethanol were recorded over a 400- hour period and the results are shown in the graph below.

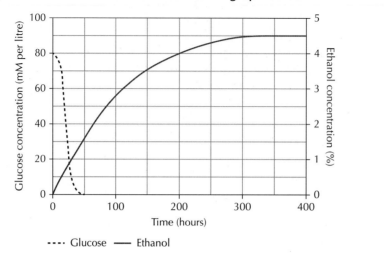

····· Glucose —— Ethanol

(a) (i) Identify the ethanol concentration in the fermenter when the glucose concentration was 75% of its starting value.

_____ % **1**

(ii) Give the time taken for the glucose to be completely removed from the solution.

_____ hours **1**

(iii) Calculate the average rate of ethanol production per hour over the first 200 hours of the investigation.

space for calculation

_____ % ethanol per hour

1

(b) Describe the evidence which suggests that the yeast takes up glucose rapidly before fermenting it more slowly.

_____

_____ **2**

(c) Give the ethanol concentration in beer made from this fermentation.

_____ % **1**

Total marks **6**

MARKS
Do not write in this margin

5. The diagram below shows a reflex arc in a human and the neurons involved.

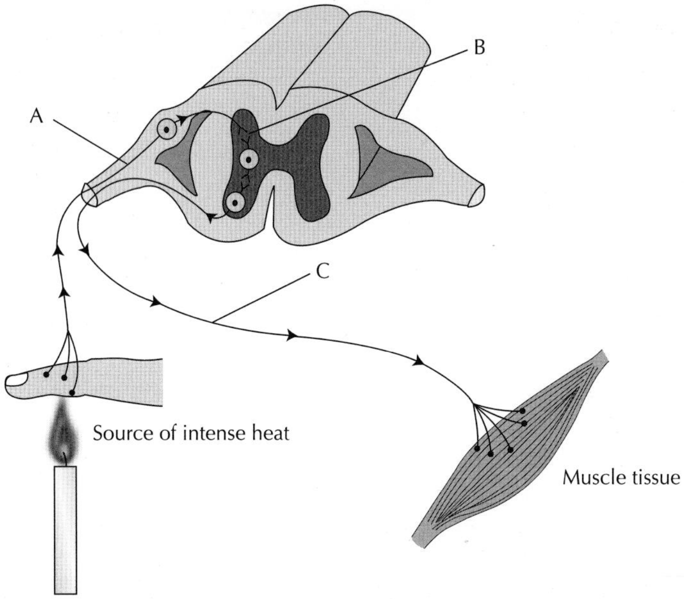
B
A
C
Source of intense heat
Muscle tissue

(a) Identify the type of neuron shown at A.

______________________________ **1**

(b) Name the gap at B and describe the role of chemicals that enter this gap.

Name __________________________

Role ____ **2**

(c) Explain the advantage of this reflex to the human involved.

____

____ **2**

Total marks 5

6. Tongue-rolling in humans is controlled by a single gene.
The dominant allele is tongue-rolling (**R**) and the recessive allele is non-rolling (**r**).

The diagram below shows the inheritance of tongue-rolling in part of a family.

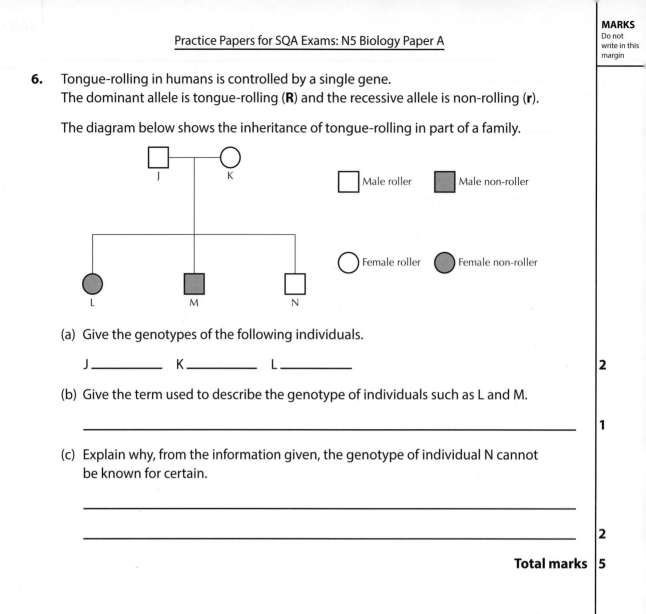

□ Male roller   ■ Male non-roller

○ Female roller   ● Female non-roller

(a) Give the genotypes of the following individuals.

J _____   K _____   L _____

|2

(b) Give the term used to describe the genotype of individuals such as L and M.

_____

|1

(c) Explain why, from the information given, the genotype of individual N cannot be known for certain.

_____

_____

|2

Total marks |5

7. The diagram below shows cells from tissues involved in the transport of substances in a plant stem.

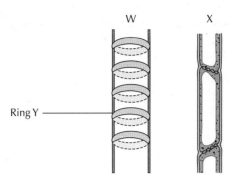

(a) Complete the table below to name tissues W and X and give **one** substance transported by each.

Tissue	Name	Substance transported
W		
X		

3

(b) Name the substance of which ring Y is composed.

_____

1

Total marks **4**

8. (a) The diagram below shows cells associated with a gas exchange surface in human lung tissue following inhalation of a breath of air.

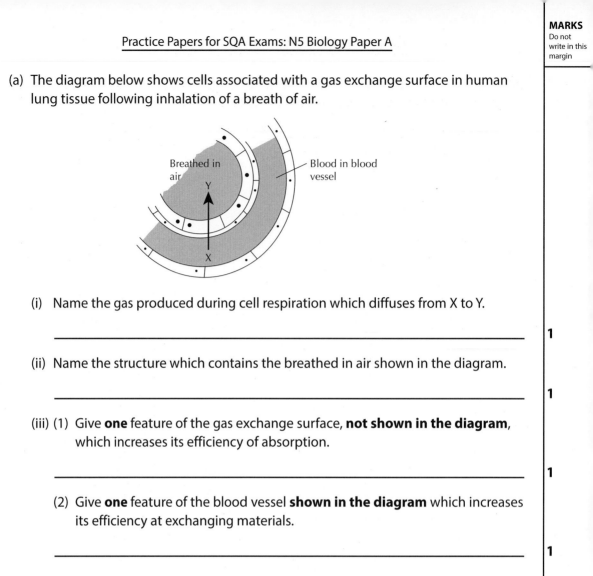

Breathed in
air

Blood in blood
vessel

Y

X

(i) Name the gas produced during cell respiration which diffuses from X to Y.

_____ **1**

(ii) Name the structure which contains the breathed in air shown in the diagram.

_____ **1**

(iii) (1) Give **one** feature of the gas exchange surface, **not shown in the diagram**, which increases its efficiency of absorption.

_____ **1**

(2) Give **one** feature of the blood vessel **shown in the diagram** which increases its efficiency at exchanging materials.

_____ **1**

(b) The chart below shows some information relating to the annual death rate of males in an area of the UK from coronary heart disease over the course of one year.

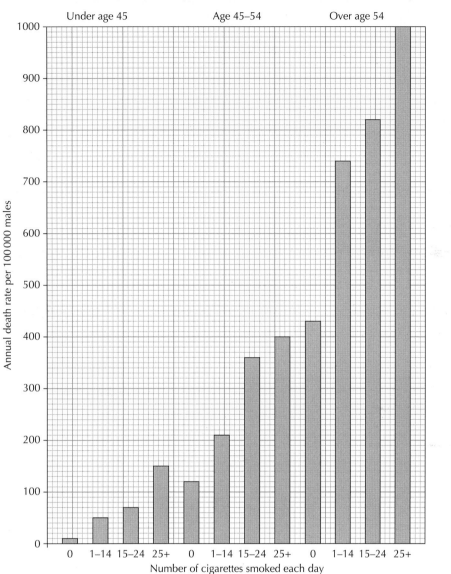

(i) From the data, identify **two** factors that affect the death rate from coronary heart disease.

1 _____

2 _____   **1**

(ii) Calculate the percentage increase in death rate in males aged under 45 years when the number of cigarettes smoked per day is increased from 1–14 to 25+.

Space for calculation

_____% **1**

Total marks | 6

9. The diagram below shows cells from a sample of human blood.

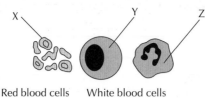

Red blood cells   White blood cells

(a) Describe the shape of cell X and explain how that shape allows it to carry out its function efficiently.

Shape _____ **1**

Explanation _____

_____ **2**

(b) Complete the table below to show the names and functions of white cells Y and Z. **2**

Cell	Name	Function
Y		produces antibodies which destroy pathogens
Z	phagocyte	

Total marks **5**

10. The bar charts below show the results of an investigation carried out to compare the numbers of four different species of ground layer plants in a hectare of woodland with the numbers found in a hectare of grassland nearby.

(1 hectare = 10 000 m²)

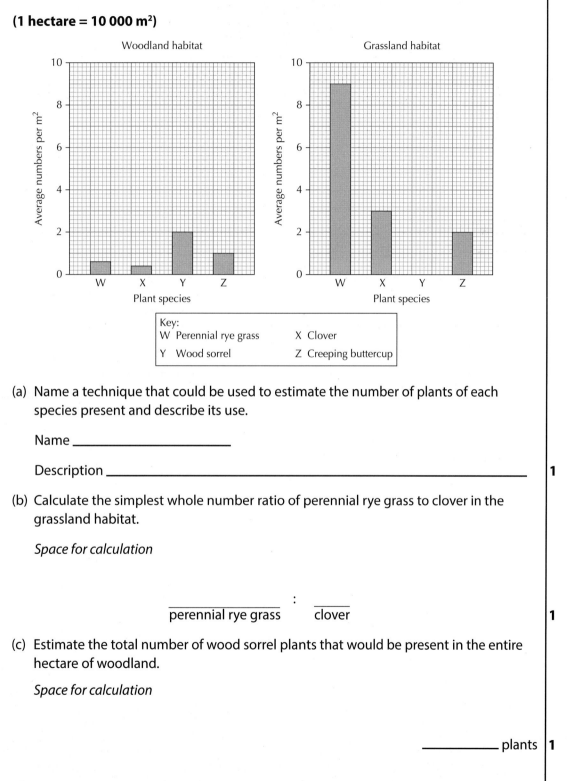

Key:
W Perennial rye grass   X Clover

Y   Wood sorrel   Z Creeping buttercup

(a) Name a technique that could be used to estimate the number of plants of each species present and describe its use.

Name _____

Description _____ **1**

(b) Calculate the simplest whole number ratio of perennial rye grass to clover in the grassland habitat.

Space for calculation

_____ : _____

perennial rye grass   clover   **1**

(c) Estimate the total number of wood sorrel plants that would be present in the entire hectare of woodland.

Space for calculation

_____ plants **1**

(d) **Choose** an abiotic factor that might be involved in the different abundance of perennial rye grass in these two habitats and explain its role.

Abiotic factor _____

Explanation _____

_____ 2

Total marks 5

11. The diagram below shows a slow-flowing stream passing through an area of farmland and five sampling sites along the stream. Fertiliser was applied to the land only in the area shown.

Six months after the fertiliser was applied, the water in the stream was sampled and its oxygen and bacterial contents measured. The results from the five sites are shown in the table below.

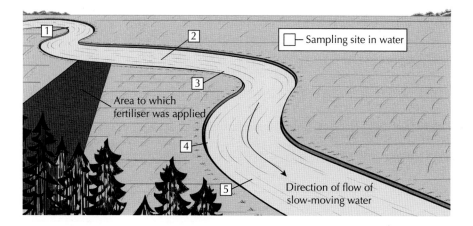

Sampling site	Oxygen level (units)	Bacterial numbers
1	140 000	low
2	400	very high
3	800	high
4	4500	high
5	16 000	low

(a) It was suggested that fertiliser from the farmland had entered the stream. Explain how this might have happened.

_____ **1**

(b) Use information from the table to describe how the fertiliser from the fields might have caused the changes in oxygen levels shown in the table between Sampling site 1 and 2 over the six months.

_____

_____

_____ **3**

(c) Identify evidence from the table which suggests that the effects of the fertiliser might not be permanent.

_____ **1**

Total marks **5**

12. Read the following passage and answer the questions based on it.

Biodiversity indicators

Terrestrial breeding birds are a good indicator of overall biodiversity. Birds respond quickly to variation in habitat quality, through changes in breeding success, survival or distribution. Since most bird species are relatively easy to identify and count and are abundant and active during daytime, they are often used as indicators of biodiversity.

Terrestrial breeding birds in Scotland include familiar garden species such as blackbird and robin, woodland species such as willow warbler and goldcrest, farmland species such as linnet and goldfinch, and species of the uplands such as raven and black grouse .

The index of numbers of terrestrial breeding birds is used as an indicator of biodiversity. The index compares bird numbers against the 1994 figure which is taken as 100. In 2007 the index stood at 121.1 and in 2015 the index was 118.2.

(a) (i) Give **two** reasons why terrestrial birds are used as indicators of biodiversity.

1 _____

2 _____ **2**

(ii) Give **one** environmental factor, not mentioned in the passage, which is monitored using indicator species.

_____ **1**

(b) Give **one** example of a bird species which breeds in farmland habitats.

_____ **1**

(c) Calculate the difference in the terrestial bird index between the following dates:

(i) 2007 and 2015

_____ **1**

(ii) 1994 and 2015

_____ **1**

Total marks **6**

13. The diagram below shows parts of two stages of photosynthesis in a green plant.

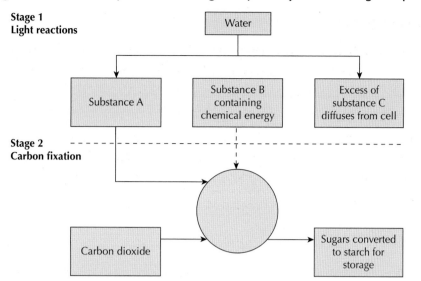

(a) Complete the table below by naming substances A, B and C produced during Stage 1.

Substance	Name
A	
B	
C	

2

(b) The experiment shown in the diagram below was used to investigate the requirements for photosynthesis in a green plant.

The plant was kept in darkness for 24 hours before being placed in bright light for 5 hours.

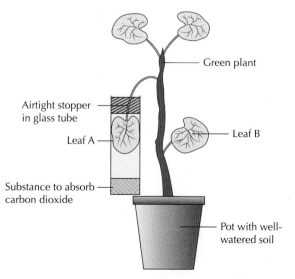

MARKS Do not write in this margin

(i) After the apparatus had been in bright light for 5 hours, a test for starch was carried out on **leaf A**.

Predict whether a positive or negative starch test would be obtained and give a valid conclusion about the requirements for photosynthesis that can be drawn from it.

Result ______________________ 1

Conclusion ______________________

______________________ 1

(ii) Describe how **leaf B** would be treated so that it could act as a control in this experiment.

____

______________________ 1

(iii) Describe how the apparatus could be altered to show that light is needed for photosynthesis.

____

______________________ 1

____

Total marks 6

14. The table and diagrams below give information about the beaks of two species of finch and a description of the habitats they occupy on the Galapagos Islands.

Size and shape of beak	Description of habitat
wide, deep and blunt	woodland with flowering shrubs providing large seeds and nuts
long, narrow and pointed	woodland with rotting logs providing food for insects

Finch species P

Finch species Q

(a) Identify the finch species that eats large seeds and give a reason for your choice.

Species _____

Reason _____

_____ **1**

(b) Suggest **two** ways in which competition between the two species is reduced.

1 _____

2 _____ **2**

(c) These two species may have arisen by evolution from a common ancestor. The processes below are involved in the formation of new species.

P   Mutation

Q   Natural selection

R   Isolation

Complete the flow chart below by adding the letters to show the order in which these processes would have occurred to produce the two species of finch.

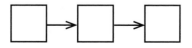
1

Total marks **4**

[END OF QUESTION PAPER]

ADDITIONAL GRAPH PAPER

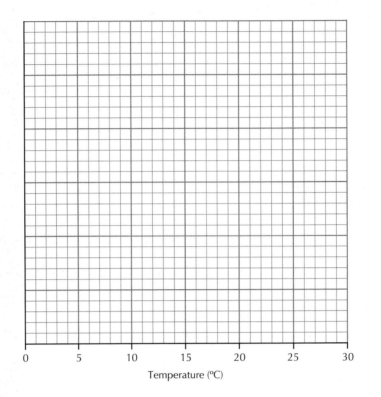

Temperature (°C)

Practice paper B

SECTION 1 ANSWER GRID

Mark the correct answer as shown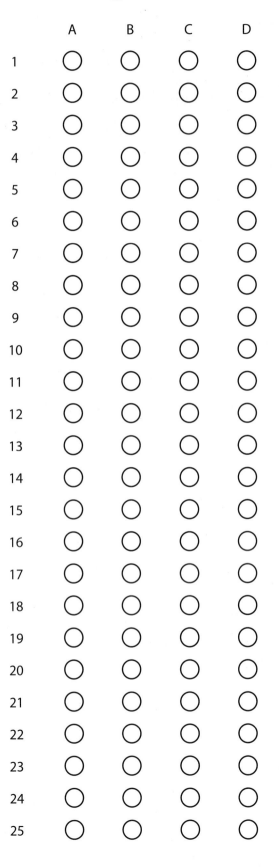

	A	B	C	D
1	○	○	○	○
2	○	○	○	○
3	○	○	○	○
4	○	○	○	○
5	○	○	○	○
6	○	○	○	○
7	○	○	○	○
8	○	○	○	○
9	○	○	○	○
10	○	○	○	○
11	○	○	○	○
12	○	○	○	○
13	○	○	○	○
14	○	○	○	○
15	○	○	○	○
16	○	○	○	○
17	○	○	○	○
18	○	○	○	○
19	○	○	○	○
20	○	○	○	○
21	○	○	○	○
22	○	○	○	○
23	○	○	○	○
24	○	○	○	○
25	○	○	○	○

N5 Biology

Practice Papers for SQA Exams

Practice Paper B
Section 1

Fill in these boxes and read what is printed below.

Full name of centre

Town

Forename(s)

Surname

Try to answer ALL of the questions in the time allowed.

You have 2 hours and 30 minutes to complete this paper.

Write your answers in the spaces provided, including all of your working.

the education publisher
for Scotland

SECTION 1 – 25 marks
Attempt ALL questions

1. The diagram below shows some structures present in a fungal cell.

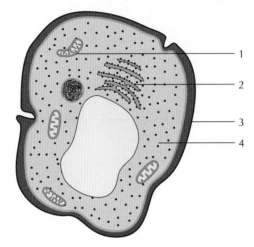

Which line in the table below identifies correctly the site of aerobic respiration and the structure that provides support for the cell?

	Site of aerobic respiration	Provides support for cell
A	1	4
B	2	3
C	2	4
D	1	3

2. The diagram below shows cells in a piece of onion epidermal tissue as seen under a microscope.

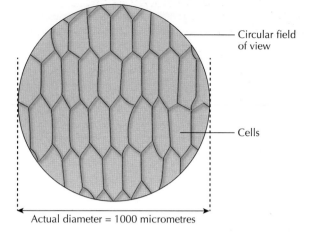

Actual diameter = 1000 micrometres

The best estimate of the average **width** of the cells shown is

A   10 micrometres

B   25 micrometres

C   100 micrometres

D   250 micrometres

3. Questions 3 and 4 refer to the diagram below which shows molecules present in a section of highly magnified cell membrane.

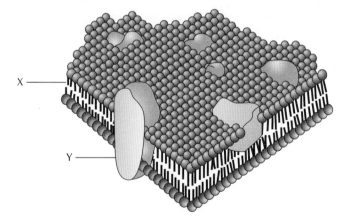

Which line in the table below correctly identifies these molecules?

	Molecule X	Molecule Y
A	protein	phospholipid
B	phospholipid	cellulose
C	protein	cellulose
D	phospholipid	protein

4. Which line in the table shows requirements for active transport?

	Energy required	*Membrane proteins required*
A	Yes	No
B	No	No
C	Yes	Yes
D	No	Yes

5. Which line in the table below shows correctly the terms that apply to the descriptions of enzyme action given?

	Description of enzyme action	
	best conditions for enzyme action	*effect of reaction on enzyme molecules*
A	optimum	denatured
B	specific	unchanged
C	optimum	unchanged
D	specific	denatured

6. The diagram below shows stages in an enzyme-catalysed synthesis reaction.

P + Q + R → S → T + R

Which of the following represents the enzyme-substrate complex?

A   P + Q + R

B   S

C   P + Q

D   T + R

7. The following are stages in the genetic engineering of bacteria.

1   Insert plasmid into host cell

2   Extract required gene from chromosome

3   Insert required gene into plasmid

4   Remove plasmid from host cell.

Which is the correct sequence of stages that would be carried out during the process of genetic modification of the bacteria?

A   2→4→1→3

B   2→4→3→1

C   3→4→1→2

D   3→1→4→2

8. The diagram below shows respiratory pathways in a mammalian muscle cell.

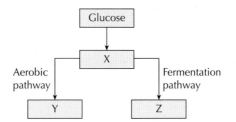

Which line in the table below identifies correctly the substances in boxes X, Y and Z?

	X	Y	Z
A	pyruvate	carbon dioxide and water	lactate
B	pyruvate	lactate	carbon dioxide and water
C	lactate	pyruvate	carbon dioxide and water
D	carbon dioxide and water	pyruvate	lactate

9. The respirometer shown below was used in an investigation of respiration in yeast.

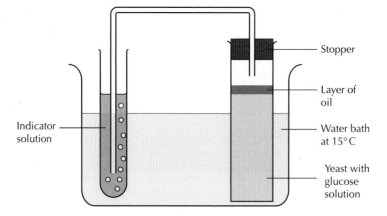

Which change to the apparatus would cause a **decrease** in the respiration rate of yeast?

A   Leaving out the oil layer

B   Diluting the glucose solution

C   Increasing the water bath temperature to 20 °C

D   Using cotton wool instead of a rubber stopper.

10. Which of the following statements about stem cells is **not** correct.

Stem cells

A   are specialised cells

B   divide by mitosis to self-renew

C   are involved in growth and repair

D   can become different types of cell.

11. A homozygous black-coated male mouse was crossed with a homozygous brown-coated female.

All the F_1 mice were black.

The F_1 mice were allowed to mate, and the F_2 generation contained both black and brown mice.

What evidence is there that the allele for black coat is dominant to the allele for brown coat?

A   Only one of the original parents was black

B   The original male parent was black

C   All of the F_1 mice were black

D   Some of the F_2 mice were black.

12. In a breeding experiment with *Drosophila*, homozygous normal winged flies were crossed with homozygous vestigial winged flies. All of the F_1 were normal winged.

If flies from the F_1 were crossed, what percentage of their offspring would be expected to have normal wings?

A 25%

B 50%

C 75%

D 100%

13. The diagram below shows a vertical section through a flower.

Which part produces male gametes?

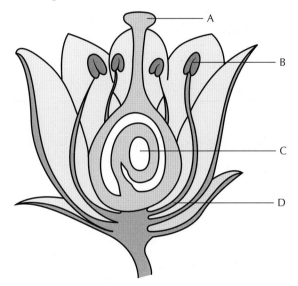

14. The diagram below shows a section through a green leaf.

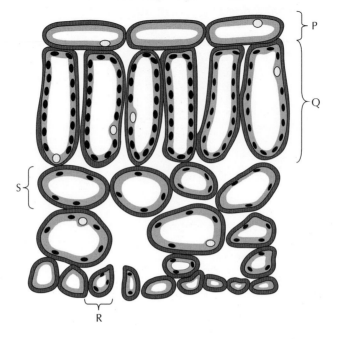

Which part of the leaf is **not** involved in the production of sugar by photosynthesis?

A P

B Q

C R

D S

15. The plant tissue that carries sugar from the leaves to the roots is the

A mesophyll

B xylem

C phloem

D epidermis.

16. The diagram below shows a single villus from the small intestine of a mammal.

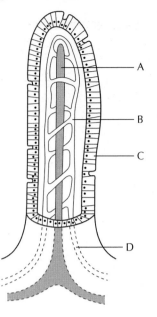

Which part is the lacteal?

17. Which line in the table below identifies correctly examples of biotic and abiotic factors affecting biodiversity?

	Biotic	Abiotic
A	temperature	grazing
B	pH	temperature
C	grazing	predation
D	predation	pH

18. The diagram below represents a pyramid of numbers.

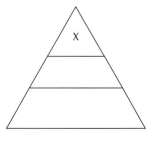

The block at X best represents the total numbers of

A producers

B herbivores

C predators

D prey.

19. Which labelled organism in the food web shown below has the **least** number of interspecific competitors for each food source?

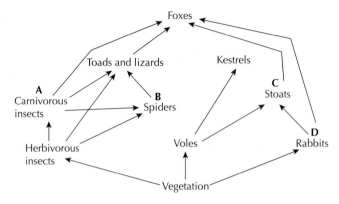

20. Various aspects of a river were sampled at five points. The results are shown in the table below.

Aspect sampled	Sampling points				
	1	2	3	4	5
Mayfly nymph number	89	15	0	0	0
Midge larvae numbers	0	1	2	175	24
Oxygen concentration (% of maximum)	85	85	75	30	63
pH level	5.5	6.0	6.4	7.3	8.0

Based on the results in the table, which of the following conclusions is valid?

A High oxygen concentration limits the numbers of midge larvae

B pH level is proportional to oxygen concentration

C Midge larvae do not survive in water with low oxygen concentration

D Mayfly numbers depend on oxygen concentration alone.

21. The apparatus below was set up to investigate photosynthesis in an aquatic plant.

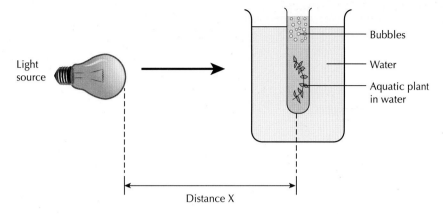

The list below shows variables related to photosynthesis that could be measured.

1   Light intensity

2   Rate of carbon fixation

3   Rate of bubble production

If distance X was increased, which variable(s) on the list would **decrease**?

A   1 only

B   2 only

C   1 and 3 only

D   1, 2 and 3

22. The role of chlorophyll in photosynthesis is to trap

A   light energy for ATP production

B   light energy for carbon dioxide absorption

C   chemical energy for carbon dioxide absorption

D   chemical energy for ATP production.

23. The following stages are involved in speciation.

   1   Natural selection

   2   Isolation

   3   Mutation

   In which order do these occur?

   A   2→3→1

   B   1→2→3

   C   2→1→3

   D   3→2→1

24. The graph below shows the increase in the human population between the years 1400 and 2000.

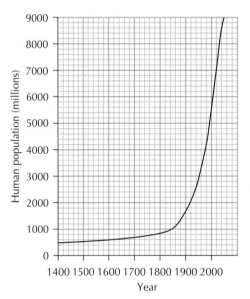

What was the percentage increase in the population between 1850 and 1950?

   A   60%

   B   200%

   C   250%

   D   50%

25. In an investigation into intraspecific competition, a batch of cress seeds of the same variety were planted out into three containers, as shown below. The containers were well watered then placed together in a bright evenly-lit room.

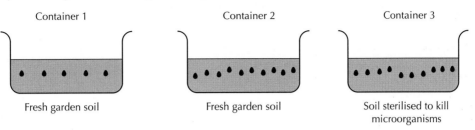

The diagrams below show the appearance of the containers after three days.

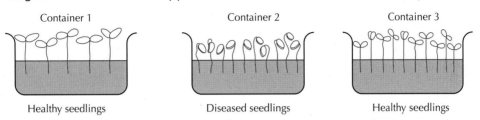

Which line in the table below correctly identifies the factor(s) involved in the diseased state of the seedlings in Container 2?

	Factors		
	sowing density	*microorganisms in soil*	*light intensity*
A	✔	✔	✔
B	✔	✕	✕
C	✕	✔	✕
D	✔	✔	✕

Key

✔ Factor involved
✕ Factor not involved

N5 Biology

Practice Papers for SQA Exams

Practice Paper B
Section 2

Fill in these boxes and read what is printed below.

Full name of centre

Town

Forename(s)

Surname

Try to answer ALL of the questions in the time allowed.

You have 2 hours and 30 minutes and thirty minutes to complete this paper.

Write your answers in the spaces provided, including all of your working.

SECTION 2 – 75 marks
Attempt ALL questions

1. Thin pieces of onion epidermis were immersed in solutions, as shown in the diagram below, and left for one hour.

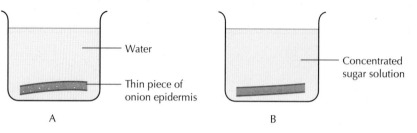

Water

Thin piece of
onion epidermis

A

Concentrated
sugar solution

B

(a) The diagram below shows the appearance of an onion cell from dish A after one hour.

Complete the diagram to show the predicted appearance of a cell from dish B after this time.

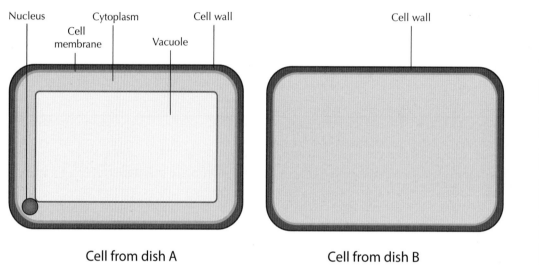

Nucleus   Cytoplasm   Cell wall

Cell
membrane   Vacuole

Cell wall

Cell from dish A   Cell from dish B

2

(b) Give the term used to describe the state of a cell, such as that from dish A, which has been immersed in pure water for one hour.

_____

1

(c) (i) Name the process that has led to the different appearances of the onion cells in dishes A and B.

_____

1

(ii) The process responsible for these changes is described as being passive. Give the meaning of the term passive in this example.

_____

_____

1

(d) Complete the following sentence by <u>underlining</u> the correct option in each choice bracket.

Onion epidermis is a(n) $\left\{ \begin{array}{c} \text{organ} \\ \text{tissue} \end{array} \right\}$ which is made up of cells carrying out a

$\left\{ \begin{array}{c} \text{similar} \\ \text{different} \end{array} \right\}$ function.

1

Total marks 6

2. The diagram below represents cells from a region of cell division in a young plant root.

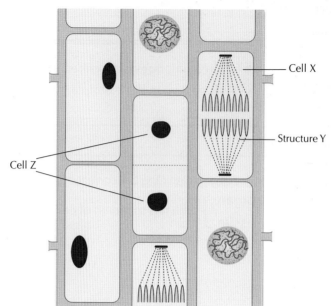

Cell X

Structure Y

Cell Z

(a) Describe the stage of mitosis shown in cell **X**.

_____

_____ | 1

(b) Name the structure labelled **Y**.

_____ | 1

(c) Complete the sentences below by <u>underlining</u> the correct option in each of the choice brackets.

The two nuclei in cell **Z** are genetically $\left\{\begin{array}{c}\text{different}\\\text{identical}\end{array}\right\}$ to each other, and each has the

$\left\{\begin{array}{c}\text{haploid}\\\text{diploid}\end{array}\right\}$ number of chromosomes. The two cells that are forming will be

$\left\{\begin{array}{c}\text{specialised}\\\text{unspecialised}\end{array}\right\}$. | 2

Total marks | 4

3. The diagram below shows a stage of protein synthesis in which messenger RNA (mRNA) is formed in the nucleus of a cell.

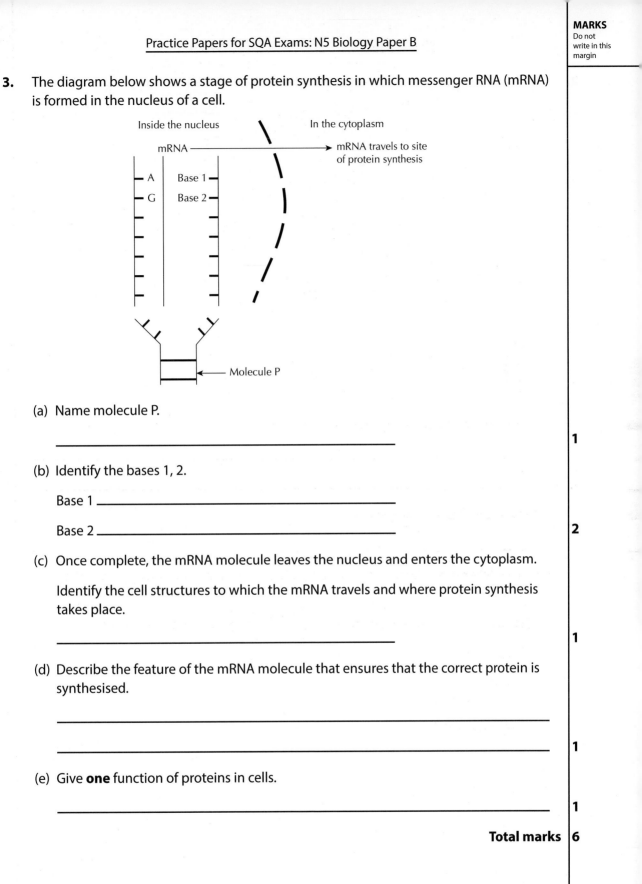

(a) Name molecule P.

_____ **1**

(b) Identify the bases 1, 2.

Base 1 _____

Base 2 _____ **2**

(c) Once complete, the mRNA molecule leaves the nucleus and enters the cytoplasm.

Identify the cell structures to which the mRNA travels and where protein synthesis takes place.

_____ **1**

(d) Describe the feature of the mRNA molecule that ensures that the correct protein is synthesised.

_____

_____ **1**

(e) Give **one** function of proteins in cells.

_____ **1**

Total marks **6**

4. An experiment was carried out to investigate the effect of temperature on the digestion of lipid (fat) by the enzyme lipase.

The experiment was set up as shown in the diagram below.

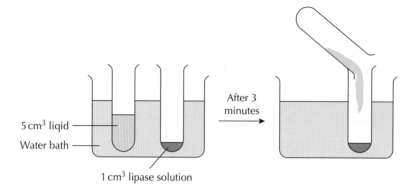

5 cm³ liqid
Water bath
After 3 minutes
1 cm³ lipase solution

The time taken for the lipid to be fully digested was recorded and the results are shown in the table below. The experiment was carried out at various temperatures and the results are shown in the table below.

Temperature (°C)	Average time taken until no lipid remained (minutes)
5	40
20	20
40	5
50	30
90	lipid remained undigested after 120 minutes

(a) Explain why the test tubes of lipid and lipase solution were kept separately in the water bath for three minutes at each temperature before mixing.

_____ 1

(b) Give **one** variable that should have been controlled to allow a valid conclusion to be made.

_____ 1

(c) Describe the effect of increasing temperature on the activity of lipase.

_____ 1

(d) Suggest **one** improvement that could increase the reliability of the results.

_____ 1

Total marks 4

5. The structure of a sperm cell is shown in the diagram below.

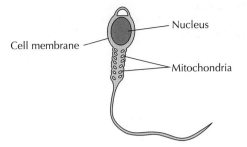

(a) Name the organs in which sperm cells are produced.

_____ 1

(b) Explain why there are a large number of mitochondria in the sperm cell.

_____ 1

(c) A sperm cell is haploid.

Explain the meaning of this statement in terms of the chromosome complement.

_____

_____ 1

(d) Describe what happens during fertilisation.

_____

_____ 1

Total marks 4

6. In an investigation of fermentation, 20 cm³ of a yeast suspension was added to 50 cm³ of grape juice and the carbon dioxide gas produced was collected and measured, as shown in the respirometer below.

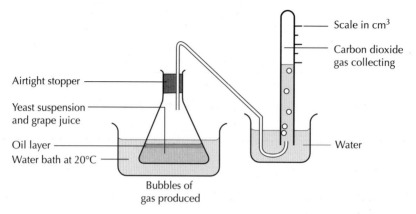

The rate of fermentation was calculated every 2 days for 10 days. The results are shown in the table below.

Day	Rate of fermentation (cm³ carbon dioxide produced per hour)
0	0
2	15
4	25
6	30
8	12
10	2

(a) On the grid below plot a line graph to show the rate of fermentation against time.

(A spare grid, if required, can be found at the end of the practice paper.)

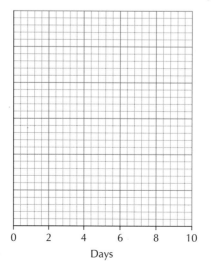

Days

2

(b) Calculate the simplest whole number ratio of volume of carbon dioxide produced per hour after 2 days to that produced after 8 days.

Space for calculation

_____ : _____

2 days   8 days

1

(c) Suggest a reason for the reduction in rate of fermentation after day 6.

_____

_____

1

(d) Suggest an improvement to the method described that would allow the investigation to be repeated more accurately.

_____

_____

1

(e) The list shows various factors that could affect the rate of respiration.

temperature   **concentration of grape juice**   **concentration of yeast suspension**

Choose a factor and describe how the apparatus could be used to investigate its effect on the rate of fermentation.

Factor chosen _____

Description _____

_____

1

Total marks   **6**

7. The diagram below shows the ends of two neurons J and M, the gaps between them and an interneuron K within the spinal cord in the central nervous system of a mammal.

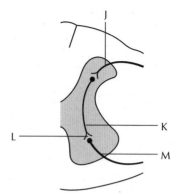

(a) Describe how a nervous message is passed along a neuron such as J.

_____ 1

(b) Describe how a nervous message arriving at the end of neuron K is able to cross the gap L.

_____

_____ 1

(c) Name gap L.

_____ 1

(d) Give **one** feature of a reflex action and explain the advantage it provides for mammals.

Feature _____ 1

Advantage _____ 1

Total marks **5**

8. The bar chart below shows variation in the length of seeds harvested from a broad bean plant.

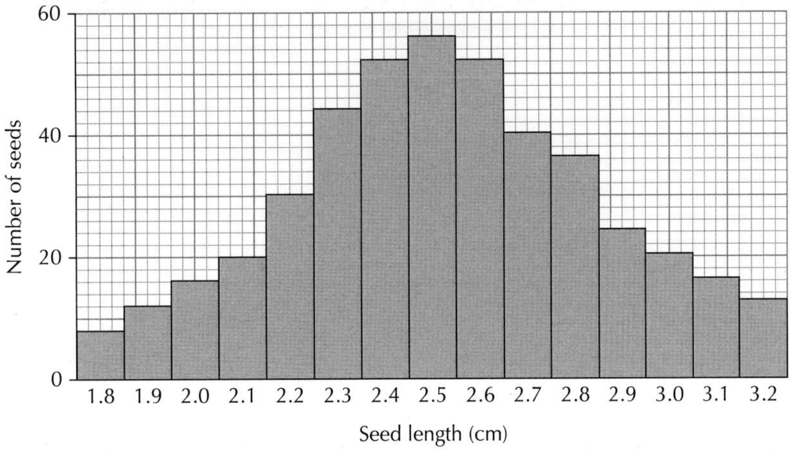

(a) Calculate the difference between the shortest and longest seeds in the sample.

Space for calculation

_____ cm | **1**

(b) Give evidence to support the statement that the seed length shows continuous variation.

_____

_____ | **1**

(c) Give **one** example of a characteristic from a **named** animal or plant species that shows discrete variation.

Named species _____

Characteristic _____ | **1**

Total marks | **3**

9. The diagram below shows the heart and an outline of the circulatory system of a human.

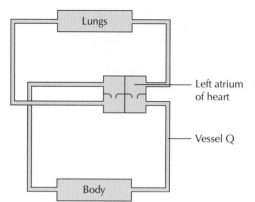

(a) **On the diagram**:

    (i)   Use the letter P to label the pulmonary artery.

    (ii)   Draw an arrow on vessel Q to show the direction of blood flow.

(b) Name the structures found in the heart and veins that prevent the backflow of blood.

_____

(c) Describe the structure of red blood cells and explain how they are adapted to take up and transport oxygen.

_____

_____

_____

_____

_____

Total marks 5

MARKS
Do not write in this margin

10. The graph below shows the average transpiration rate of barley plants in an open field over a 24-hour period during summer in Scotland.

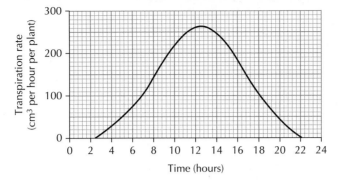

(a) Give the period during which the average transpiration rate is greater than 100 cm³ per hour per plant.

From _____ hours until _____ hours

1

(b) Name **two** environmental factors that might be involved in the changing rates of transpiration over the period.

1 _____

2 _____

2

(c) The diagram below shows cells in the lower epidermis of a barley leaf.

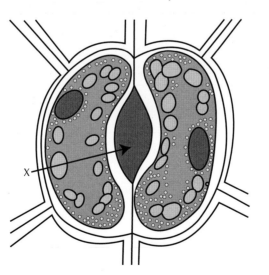

Name structure X, through which water vapour leaves the plant during transpiration.

_____

1

(d) Suggest a possible benefit of transpiration to a barley plant.

_____

1

Total marks **5**

11. The charts below show the occurrence of five species of plants in samples taken from an area of grassland and from a path passing through it.

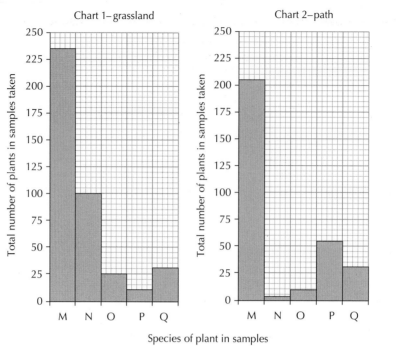

Chart 1– grassland

Chart 2–path

Species of plant in samples

(a) Name **one** method that could be used to sample plants in grassland and describe its use.

Name _____

1

Description of use _____

_____

1

(b) Give the number of plants of species M that were found in samples taken from the path.

_____ plants

1

(c) Describe the effects on the numbers of species O and P of being walked over by people using the path.

Species O _____

1

Species P _____

1

(d) Give the species that is least affected by being walked over.

Species _____

1

Total marks 6

12. A group of students set four pitfall traps in a woodland to sample the leaf litter invertebrates living there. The traps were left set for the same length of time.

The table below shows the number and types of invertebrates found.

Invertebrates	Number
Woodlice	50
Snails	5
Centipedes	15
Beetles	35

(a) Use the information in the table to complete the pie chart below.

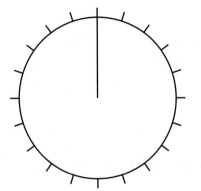

2

(b) Give **one** source of error which can arise when setting up a pitfall trap and suggest **one** method of minimising it.

Source of error _____

_____ 1

Method of minimising error _____

_____ 1

Total marks 4

13. The graph below shows the change in plant growth per hectare when different masses of nitrate were added to fields growing an identical variety of crop plant.

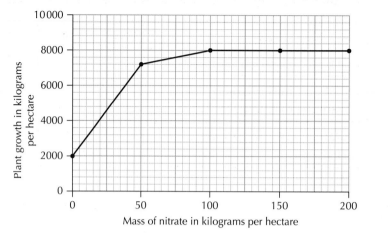

(a) Give the mass at which the concentration of nitrate ceases to be a limiting factor in the growth of the crop plant.

_____ kilograms per hectare **1**

(b) Calculate the percentage increase in plant growth when the mass of nitrate is increased from 0 to 100 kilograms per hectare.

Space for calculation

_____% **1**

(c) Name **one** type of substance that is produced by plants using nitrate absorbed from the soil.

_____ **1**

(d) Name the nitrate-containing substances that are added to soil by farmers to increase the yield of their crops.

_____ **1**

Total marks **4**

MARKS
Do not write in this margin

14. (a) The table below refers to mutation.

Decide if each statement in the table is true or false and tick (✔) the appropriate box.

Statement	True	False
Mutation is a non-random event.		
Mutation can confer an advantage to an organism.		
Mutation is the only source of new alleles.		

2

(b) Describe how natural selection is involved in the evolution of new species.

_____

_____

_____ **3**

Total marks **5**

15. Lichens live on the surfaces of walls and trees, and are sensitive to sulfur dioxide, a gas linked with air pollution.

The graph below shows the results of a study in which the percentage of surfaces that were covered by lichens along a line from the centre of a large city that had air polluted by sulfur dioxide was estimated.

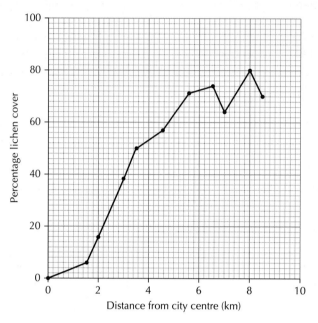

(a) Lichens can be used as indicators of air pollution.

Describe how this statement is supported by the data shown in the graph.

_____

_____ **1**

(b) At what distance from the city centre was the air pollution the lowest as indicated by percentage lichen cover?

_____ km **1**

(c) As well as sulfur dioxide, polluted air often contains tiny black soot particles.

Predict how these particles would affect the rate of photosynthesis in plants growing in polluted air. Explain your answer.

Prediction _____ **1**

Explanation _____

_____ **1**

Total marks **4**

16. Read the following passage and answer the questions based on it.

Blood oxygen levels during exercise

A study was carried out to compare the oxygen saturation of blood taken from muscles in different states of activity. The states of activity included lying down, walking on a treadmill at 1 mile per hour, and jogging on a treadmill at 5 miles per hour.

In order to measure oxygen saturation, catheters with an oxygen sensor were inserted into the left femoral artery and left femoral vein, which transport arterial and venous blood to and from the left leg, respectively. The catheters continuously sampled the blood and measured the oxygen saturation of the blood.

Oxygen saturation is equal to how much oxygen the haemoglobin in the blood is carrying, as a percentage of how much it could theoretically carry. A higher oxygen saturation value implies a higher concentration of oxygen in the blood.

The measurements of the oxygen saturation in the blood taken from the leg muscles in different states of activity are shown in the table below.

Oxygen saturation of blood sample (%)	Activity		
	Lying down	Walking	Running
Femoral artery	97	96	97
Femoral vein	81	68	65

(a) Calculate how many times greater the decrease in oxygen saturation was after running compared to lying down.

Space for calculation

_____

1

(b) The femoral artery transports oxygenated blood to the muscle cells in the legs. Explain why there is a greater decrease in the oxygen saturation in the blood of the femoral vein after running.

_____

_____

1

(c) Give **one** structural difference between an artery and a vein.

_____

_____

1

MARKS
Do not write in this margin

(d) Name the blood vessel through which gas exchange takes place between the blood and the leg muscles.

_____

1

Total marks 4

[END OF QUESTION PAPER]

ADDITIONAL GRAPH PAPER

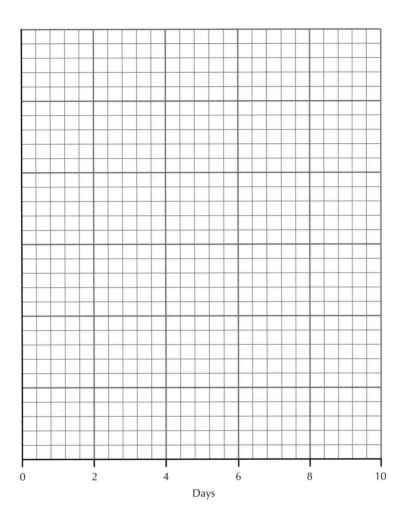

Days